SECO...

TRANSISTOR
EQUIVALENTS
AND
SUBSTITUTES

BY

B. B. BABANI

BERNARD BABANI (publishing) LTD
The Grampians
Shepherds Bush Road
London W6 7NF
England.

Although every care is taken with the preparation
of this book, the publishers will not be responsible
for any errors that might occur.

I.S.B.N. 0 85934 013 9

© 1974 by Bernard B. Babani

First published 1974
Reprinted February 1974
Reprinted August 1974
Reprinted January 1975
Reprinted October 1975
Reprinted February 1976
Reprinted October 1976
Reprinted May 1977
Reprinted April 1978
Reprinted April 1979

Without the help of my Secretary, Miss Pauline
Pragnell, my labour would have been very much
heavier.

B. B. BABANI
LONDON – 1974

Printed and Manufactured in Great Britain
by C. Nicholls & Company Ltd. Manchester

TRANSISTOR VERGLEICHSTABELLE

TABELLE DI COMPARAZIONE
DI TRANSISTORI

TABLES D'EQUIVALENCE TRANSISTORS

TRANSISTOREN VERGELIJKINGSTABELLEN

TRANSISTORES-EQUIVALENCIAS

TRANSISTORES Y REEMPLAZAOS

INTRODUCTION

Up to now, to the best knowledge of the author, there has been no book available at a reasonable price to assist Service Engineers, Home Constructors, Designers and Amateurs in determining the best available transistors to replace a given type.

This need was provided for in my First Book of Transistor Equivalents and Substitutes published in 1971 and reprinted at least 10 times since that date with more than one million copies having been sold.

At least 36,000 new types of transistors have been introduced in the last two years and it is because of this that I have prepared this the Second Book of Transistor Equivalents and Substitutes. This book contains no material that appeared in the First Book - BP1 - but includes all new transistors and their equivalents that have appeared since 1971. For complete coverage of this important field, both books are needed.

It will be realised by all users of transistors, that one cannot guarantee absolute equivalents as in the case of many electronic valves. Therefore, the only way to handle the replacement problem, is by careful consideration being given to every characteristic of both the required transistor and its possible alternatives. This has been done in the preparation of this book.

It should be carefully noted however, that in many cases the possible alternatives may be different in appearance or size and when replacing, carefully check if space is available to take the replacement.

It is also recommended that where high frequency transistor circuitry is involved and when replacements are being used, slight re-alignment may be necessary. This is also advised even if the replacement transistor is of identical type number to the one that is being replaced. In extreme cases, neutralisation may have to be provided even when unnecessary in the original circuit. This is due in the main, to comparatively large ranges of inter-electrode capacitances normally found in transistors.

As this book contains many thousands of transistors, it has been necessary to use some form of abbreviation in listing the possible replacements for a required transistor type.

For example, in the first page of the book, if one looks at the alternatives for type 2G302, the replacement types shown read as follows:

$$AF185.2N1305-1307.SK3004$$

This means that for the type 2G302, one can use as an alternative the AF185 or 2N1305 or 2N1307 or SK3004 respectively, there being no particular preference indicated in the order that the replacements are listed.

It should be noted that in the first column, the listing is shown in numerical/alphabetical order.

Included in this book are transistors of Pro-Electron, British, German, Dutch, U.S.A., Japanese, Czech, Polish and European manufacture. Thus, virtually, all known types that have possible replacements have been included.

The author has taken every care possible in the compilation of the data in this book, but neither the author nor the publishers can accept any responsibility for any errors that may be present.

B. B. BABANI
LONDON - JANUARY 1974

Transistors detailed in this book include products of the following manufacturers :-

AEG	Intermetall	Semiconductors
AEL	IRC	SES
Akers	ITT	Sesco
AMD	LCT	SGS
Amperex	LRC	Sharp
Amroh	LRT	Shindengen
Archer	LTT	Siemens
ASC	Lucas	Signetics
Associated	Mallory	Silec
Ates	Matsushita	Silicon T C
BBC	MBLE	Siliconix
Belvu	Meco	Solid State
Bendix	Micro	Solitron
Bre	Microstate	Sony
Bradley	Microwave	Sperry
BTH	Miniwatt	Sprague
CBS	Mistral	STC
Clevite	Mitsubishi	Sylvania
Continental	Motorola	Tadiran
Cosem	Mullard	Tekade
Ditratherm	National	Telefunken
Delco	NEC	Tesla
Ebauches	Newmarket	Texas
Eberle	Nortron	Thorn
ECC	Pacific	TI
Fairchild	Philco-Ford	Toshiba
Fanon	Philips	Transitron
Ferranti	Piher	Trans-Sil
FTR	Plessey	TRW
GE	Raytheon	Tung-Sol
GEC	RCA	Tungsram
GI	RFT	Valvo
GTC	RIZ	Western
Hitachi	RTC	Westinghouse
Hughes	Sanyo	
Industro	Semicon	

COLOUR CODE ABBREVIATIONS USED ON TRANSISTORS

B	Black	Zwart	Noir	Schwarz	Nero	Negro	Preto
BL	Blue	Blauw	Bleu	Blau	Azzurro	Azul	Azul
BR	Brown	Bruin	Brun	Braun	Bruno	Moreno	Castanho
G & GP	Green	Groen	Vert	Grun	Verde	Verde	Verde
OR	Orange	Oranje	Orange	Orange	Arancia	Naranja	Laranja
R	Red	Rood	Rouge	Rot	Rosso	Rojo	Encarnado
V	Violet	Violet	Violet	Violett	Violetta	Violeta	Violeta
W	White	Wit	Blanc	Weiss	Bianco	Blanco	Branco
Y	Yellow	Geel	Jaune	Gelb	Giallo	Amarillo	Amarelo

EQUIVALENTS OF U.S.A. CONSUMER TYPE TRANSISTORS.

TANDY-ARCHER	DIRECT	R.C.A. & U.S.A.	PRO-ELECTRON	G.E.
RS276-2001	2N1304	SK3010	AC127	HEP641
RS276-2002	2SA221		AC130	HEP3
RS276-2003	2SD187		AC107	
RS276-2094	2SB22 Or	SK3005	ACY19	HEP252
RS276-2005	2SB322Y	SK3003	AC109	HEP254
RS276-2006	2SD407	SK3004	AD103	HEP200
RS276-2007	2N1305	SK3009	ASY26	HEP629
RS276-2008	2N5058			
RS276-2009	2N2222	SK3045	BC107	HEP50
RS276-2010	2N2484	SK3020	BCY59	
RS276-2011	2N918	SK3018	BFY27	HEP56
RS276-2012	A5T5058			
RS276-2013	TIS97		BC14T	HEP726
RS276-2014	2N5449			
RS276-2015	2N1996			
RS276-2016	A5T3904	2N4286 /7	BC113	HEP720
RS276-2017	SP8918	2N4910 /2		HEP54
RS276-2018	T1P29	SK3024	BLY61	
RS276-2019	T1P33			HEP 243
RS276-2020	T1P3055			
RS276-2021	2N2907		BC143	HEP51
RS276-2022	2N2605		BC177A	
RS276-2023	A5T2907	2N5058 /7	BC116	HEP52
RS276-2024	2N5447		BC154	HEP57
RS276-2025	SP8919	SK3025		HEP242
RS276-2026	T1P30	2N4919	BD138	HEP246
RS276-2027	T1P34	2N4901/2		HEP700
RS276-2028	A5T3821	TIS14	BC178B	HEP801
RS276-2029	2N4891			

2G101/2	AC126-163-171-173, 2SB459, 2N506-2907, AF139, NKT677F.
2G106	AF106-192-256, BFX48, BSW19-72, NKT674F, 2N2273-2873-4916-5354.
2G108/9	AC126-163-171-173, 2SB459, 2N2429-2907.
2G110	AF106-192-256, BFX48, BSW19-72, NKT674F, 2N2273-2873-4916-5354.
2G138	AF185, SK3005.
2G139	AF106-185, SK3005.
2G140	AF185, SK3005.
2G141	AF106-185.
2G201/2	AC117R-128-153, 2SB370, 2N2431-4106.
2G240	AU103-105, 2SB468, SK3014.
2G270/1	AC114-128-132-153, 2SB222, 2N2431-4106, SK3004.
2G301	AC114-128-153, AF185, 2SB222, 2N1305-2431-4106.
2G302	AF185, 2N1305-1307, SK3004.
2G303/4	2N1305-1307.
2G306	2N1309, SK3004.
2G308	2N1305-1307, SK3004.
2G309	2N1309.
2G319/20	AC126-128-163-171, 2SB383, 2N1190-2429, SK3005.
2G339	AC127, 2SD96, 2N2430, SK3010.
2G371	AC117K-121V-128-131-132-152V-184, 2S32, 2N610-1305, OC318, NKT213, SK3005.
2G374	AC121VII-128-131-132-152V-184, 2SB415, 2N1309, OC318, SK3005.
2G377	NKT217.
2G381	AC117K-128-153-180, 2SB222, 2N467-1305, SK3004.
2G382	AC117K-128-153-158, 2SB370, 2N1305-1373-4106.
2G383	ACY17, SK3005.
2G384	ACY18, SK3004.
2G385	ACY19, SK3004.
2G386	ACY18-20, SK3004.
2G387	ACY21, SK3004.
2G401/2	AF106-127-185-190, 2SA230.
2G415	AF124.
2G417	AF127.
2G526	ACY24K, ASY48-80-81, 2SB89, 2N527, SK3005.
2G577	ACY24K, ASY48-77-81, 2SB89, 2N241, SK3005.
2G1024/5/6	ACY24K, ASY48-77-81, 2SB89, 2N241, SK3006.

2N24A	ACY24K, ASY48-77.
2N34	AC122R-125-151-151IV, 2N1191-2429.
2N34A	AC122R-126-151IV-163-171, 2N1191-2429.
2N35	AC179-185-186, 2N2430.
2N35A	AC127-130, 2N2430, 2SD11.
2N36	AC122R-125-151IV, 2N1191-2429-2431.
2N37	AC125-151IV, OC304, 2N1191-2429-2431.
2N38	AC122R-151IV, OC304, 2N1191-2429-2431.
2N38A	AC151IV, 2N1191-2429-2431.
2N40	AC162-170, 2N706A-1191.
2N41	AC125-151, 2N404-1190-2447, 2SB459.
2N42	AC125-151, 2N404-1190-2447, 2SB459.
2N43/A	AC131/30-132-184, OC318, 2N525.
2N44	AC131/30-132-184, OC318, 2N524-2431.
2N45	AC122/30-131/30-151, 2N404-524.
2N46	2N706A.
2N47	AC122R-125-151IV, 2N404-1191.
2N48	AC122R, OC304, 2N404-1191-2431.
2N49	AC122R-151IV, OC304, 2N404-1191-2431.
2N50	AC122-125-151, 2N404-1191.
2N51	AC122-122R-125-151, OC304, 2N404-1191.
2N52	AC122-151, 2N404-1191, 2SB222.
2N53	AC151, 2N404-1191.
2N54	AC122-122R-125-151, OC304, 2N1191-2429-2431.
2N55/56	AC122-122R-151IV, OC304, 2N404-1191.
2N57	AC122-151IV, 2N1191, TI3027.
2N59/ABC	AC122G-122/30-126-151IV-VI, 2N1193, 2SB32.
2N60	AC122G-126-151VI, OC304, 2N404-1193.
2N60A	AC122-122/30-126-151IV-VI, 2N61A-1193-6147, 2SB222.
2N60B	AC122-122/30-126-151IV-VI, 2N1193-6147, 2SA219.
2N61	AC122-122Y-125-126-128-131-151-151IV-V, OC304, 2N1192-2431.
2N61A	AC122-125-126-131/30-151IV-V, 2N60-524-1192.
2N61B	AC126-131/30-151IV-V, 2N60-61-1192.
2N61C	AC122-125-131/30-151IV-163, ACY33, 2N60-61-1192, 2SB220.
2N62	AC122-122R-125-131/30-151IV, OC304, 2N109-403-1191.
2N63	AC122R-125-151IV, OC304, 2N1191-2429.
2N64	AC122R-125-151IV, 2N1191-2429.
2N65	AC122Y-126-151IV-VI, 2N1193-2429.
2N66	AD148-155-162, 2N643, TI3027.
2N68	AD139-148-155-162, TI3027.
2N69	AC122-125, 2N1190.
2N71	AC122-125, 2N404-1190.
2N72	AC122-125, 2N1190.
2N73	AC122-125, 2N1190.
2N74	AC122-125, 2N1190.
2N75	AC125, ACY24K, 2N1190.
2N76	AC162-170, 2N1191.
2N77	AC151, 2N465-565.
2N78/A	ASY29-70-80, 2N1302.
2N79	AC128-171-173, 2N1191-2429.

2N80	AC122-128-171-173, 2N2429.
2N81	AC125-128-171-173, 2N1191-2429.
2N82	AC128-171-173, 2N1191-2429.
2N83	AD148V, CTP1104, LT5036.
2N83A	AD148V, OC30, OD603, 2SB240.
2N84	AC124-128K-153, 2N109-4106.
2N85	AC131-131/30-184, BSX19, 2N1924, BSY63.
2N86	AC124R-131-153-184, 2N2429-4106.
2N87	AC124R-153, 2N2431-4106.
2N88	AC129V, AFY69, 2N109.
2N89	AC129V.
2N90	OC66.
2N93	AC162-170-173.
2N94	ASY26-2N1304.
2N94A	ASY26-74, OC140.
2N95	LT5210.
2N96	AC128-171-173.
2N97	AC127-179, ASY28, 2N169A-1302-2430.
2N97A	AC127-79, 2N169A-438-2450.
2N98	ASY26, 2SA155, 2N1304.
2N98A	ASY26-73. 2N169A-444.
2N99	ASY26, 2N169A-1306.
2N100	ASY73, 2N1306.
2N101	AD139-155.
2N102	ASY28, 2N1302, LT5210.
2N103	ASY28, 2N650-1302.
2N104	AC124R-128, OC304, 2N650-4106.
2N105	AC151-162, 2N650-1191.
2N106	AC122R-124R-128-151IV-153, OC304, 2N1191-4106.
2N107	AC122R-151IV-170, OC304, 2N1191.
2N108	AC125-170.
2N109	AC117K-122R-125-151IV-153K, OC304, 2N187A-217-1191.
2N110	AC163-171, 2N2429, 2SB383.
2N111	AF106-126-132-136-196, HS23D, 2N147.
2N111A	AF126-132-136-196.
2N112	AF132-136-137-196, 2N579.
2N112A	ASY26, AF126-132-136-196, 2N118.
2N113	AF106-115-127-132-136-190-196, 2N416.
2N114	AF132-136-137-196, 2N417.
2N115	AD130III-149-150-179, TI3029.
2N116	AC129V, OC57, 2N133A.
2N117	2N4264.
2N118	BD130, 2N118A-3055-4264.
2N118A	BD130, BDY10, MPS834, 2N3055.
2N119	BD130, 2N3055-4264.
2N120	BD130, MPS3704, 2N3055.
2N123	AF106, 2N1303.
2N124	ASY26, 2N1302.
2N125	ASY27-74, 2N1304.
2N126	ASY27.
2N127	ASY28, 2N1304.
2N128	AF126-135.
2N129	AF135.

2N130	AC122R-125R-128-151IV-162, OC304, 2N1191-1924.
2N130A	AC122/30-131-152-162, ASY48, 2SB224, 2N402-650-1413-1924.
2N131	AC122-122Y-151V, OC304, 2N569-1192.
2N131A	AC122-125-131/30-153, ASY48, 2N568-651, 2SB103.
2N132	AC122-122Y-125-151-151V, OC304, 2N446-1190-1192.
2N132A	AC125-151, 2N641-651-1190-1413, 2SB32.
2N133	AC122-122Y-125-151-151V, OC304, 2N1190-1192.
2N133A	2N651.
2N135	AF126-132-136-196, AFY15.
2N136	AF106-126-132-136-190-196.
2N137	AF106-127-132-136-190-196.
2N138	AC128-163-171, OC305, 2N4106.
2N138A	2N1008-1128-4106.
2N138B	AC117-128-153, OC318, 2N270-4106, 2SA219.
2N139	AF106-126-136-190-196, AFY15, 2N1638, 2SA36.
2N140	AF125-131-135-195, AFY15, SFT354, 2N1639.
2N141	AD139-148V.
2N143	2N1038.
2N145	2N1302.
2N146	2N1302, AUY19.
2N147TL	AUY19.
2N155	AD133-138-148-149, TI3027, 2N301, 2SB107.
2N156	AD133-138, TI3027, 2N176.
2N157	AD133-138-150IV, 2N1022A-1038-1530-1531.
2N157A	AD131-138/50, OC26-30-35, OD603, 2N1014-1532.
2N158	AD131-138-166, OC35, TI156-3027, 2N1532-2139.
2N158A	AD131-138, OC35, TI3027, 2N1532-2141.
2N160/A	MPS3706.
2N161/A	MPS3706.
2N162/A	MPS3393.
2N163/A	MPS3393.
2N166	ASY26-73, 2N1304.
2N167	2N1306.
2N168	ASY26, 2N1306.
2N169	AC127, ASY29-74, 2N1308.
2N169A	ASY29-74-75, 2N78.
2N170	ASY27, 2N1308.
2N172	ASY27, 2N1308, 2SA206.
2N173	AD138/50, TI3027.
2N174	AUY21-28, OC35, TI3027.
2N174A	ADZ12, TI3027.
2N175	AC122Y-125-151V, 2N1192-2613.
2N176	AD130-166, TI3027, 2N2869.
2N178	AD166, TI3027.
2N180	AC122Y-125-128-151V, OC304, 2N109-1192.
2N181	AC122Y-125-131-151V-162, 2N1192.
2N182	ASY27, 2N1302.
2N183	ASY27, 2N1302.
2N184	2N1306.
2N185	AC122-122R-125-151IV-153, OC304, 2N650-1191-2430.
2N186	AC122-122R-125-151IV-153, OC304, 2N1191-4106.
2N186A	AC125-128-131-151, 2N270-1191.

2N187	AC122-122R-125-151IV, OC304, 2N319-1191-4106.
2N187A	AC117-128-153, 2N270-1191, 2S37.
2N188	AC122-122R-125-151IV, OC304, 2N1191-2430.
2N188A	AC131-132-152-162-184, 2N320-1191.
2N189	AC131-151IV, 2N270-1191, 2SB219-222.
2N190	AC122-131-151-151IV, 2N322-1191.
2N191	AC131-151V-153, 2N323-1190-1192.
2N192	AC122G-131-151VI, 2N324-1191-1193.
2N193	AC122, 2N1302.
2N194	AC122, 2N1302.
2N194A	ASY73.
2N195	AC131.
2N196	AC131, 2N4106.
2N197	AC131, 2N4106.
2N198	AC131-163, 2N4106.
2N199	AC131-163, 2N4106.
2N200	AC126-128-131-153, OC71, 2N2429.
2N204	AC128-131-138-151V, OC71, 2N1373-1377.
2N205	AC128-131-138-151V, OC71, 2N1372-1377.
2N206	AC122G-126-151VI-153, OC41, 2N1193-1372-1377.
2N207	AC122R-125-151IV, 2N1191-1372-1377.
2N207A	AC122-122G-125-126-151-151VI, OC304, 2N535A-1193-1372, 2SB221.
2N207B	AC107-122-122G-125-126-132-150-151VI-163, OC304, 2N535B-1193.
2N211	AF190, 2N1302.
2N212	2N1302.
2N213	2N1304.
2N213A	AC127.
2N214	AC125-185-186, 2N1192-1304.
2N215	AC122-125-128-151, OC304, 2N217-404-1189.
2N216	AC127-128, 2N1302.
2N217	AC128-131/30-151V-152-180, 2N404-1192-4106.
2N218	AC121IV, AF126-132-136-196, AFY16, 2N204-404. 1638.
2N219	AC121V, AF126-132-136-197, SFT307, 2N140-404-1638.
2N222	AC126, 2N2429-2439.
2N223	AC125-131-151, 2N404-2431-4106.
2N224	AC125-131-151-184, 2N404-1192-4106.
2N225	AC106-117-128-131-153, 2N224-404-1193-4106.
2N226	AC125-131-151-184, OC308, 2N404-1406-2431.
2N227	AC117-128-153, OC308, 2N404-1406.
2N228	AC128-153.
2N229	AC127, 2N1302-2430.
2N230	AD150, 2N2836.
2N231	AF132-136-137-196, 2N139-404.
2N232	AF136-137-196, OC45, 2N139-404.
2N233	AC127, 2N1306.
2N233A	AC130.
2N234	AD149IV, OC26, TI156-3027, 2N554.
2N234A	AD149-149V-150, OC26, TI3027, 2N301-555.
2N235	AD131, OC26, TI3027, 2N3611.
2N235A	OC26, TI3027, 2N3611, AD131, AUY19.

2N235B	AD131-138-149, OC26, TI3027, 2N3613.
2N236	AD131, TI3027, 2SB250A.
2N236A	AD131-149, OC26, TI3027, 2N301, AUY19.
2N236B	AD131-149, OC26-30, TI3027, 2N1078, 2SB107A.
2N237	AC128, AD131-149, 2N404.
2N238	AC122-122R-125-151IV, 2N404-1191-4106.
2N239	AC117-122-122R-125-151, 2N1191-4106.
2N240	AC122R, 2N404.
2N241	AC125-131-151-163-184, 2N404-4106.
2N241A	AC125-131-151, 2N321.
2N242	AC128-131-163, 2N281-4106, 2SB221-248A.
2N243	BSY45.
2N247	AF131-135-136-195, 2N2188.
2N248	AF131-135-136-195, 2N2188.
2N250	AD130V-138, 2N250-456A-3611.
2N251	AD138/50-149, AUY22III, 2N301A-456A-1530, 2SB201.
2N251A	2N251-456A.
2N252	AF106, 2N2188.
2N253	2N1302.
2N254	2N1302.
2N255	AD148, TI3027, 2N554.
2N256	AD148, TI3027.
2N256A	AD148-149, TI3027.
2N257	AD138-150IV/V, TI3027, 2N3611, 2SB-107A.
2N258	MPS3703.
2N259	MPS3703.
2N260/A	MPS3703.
2N261	MPSA55.
2N262/A	MPS6522.
2N263	MPS3704.
2N264	MPS3394.
2N265	AC125-128-151, 2N1175.
2N266	AC117-128-151-153, 2N408.
2N267	AF131-135-136-195, 2N2188.
2N268	AD132II, AUY28, TI3027, 2N1530.
2N268A	TI3027.
2N269	ASY14, ASZ15, TI3027, 2N1303.
2N270	AC122G-128-131-151VI, 2N404-1193-4106.
2N271/A	GALLEY FOUR
2N272	AF126-132-136-196, AFY15, ASY27, 2N2905.
2N273	AC117-117R-153, BC261, GFT32, 2N4106.
2N274	AC128, 2N4106.
2N276	2N2188.
2N277	AF126-132-137-196.
2N278	AD133-138, AUY31.
2N279	AD133-138/50, AUY31.
2N280	AC122/30-128-151IV, 2N119-650-2431-4106.
2N281	AC122/30-125-128-132-153, 2N651-2431-2907-4106, 2SB221.
2N282	AC117-122/30-125-128-132-153, 2N4106.
2N283	AC122-125-128-151, 2N650-4106, 2SB225.
2N284	AC122-122/30-125-151, BC212.
2N284A	AC122/30, ASY48-48IV-76.

2N285	AD130-138-149, AUY22Ш-28, TI3027, 2N3617.
2N290	AD149, 2N2836.
2N291	AC131, 2N4106.
2N292	ASY26-73, 2N313-1302.
2N293	2N1304.
2N296	ASZ15, AUY22, 2N379-456A-3146.
2N297	AD131-131IV-138/50, TI3027, 2N297A-456A-3146, 2SB107.
2N297A	AD131-131IV-138/50, ASZ15, 2N297-457-2188.
2N299	AF124-194, 2N2188.
2N300	AF194, 2N2188-2495.
2N301	AD131-138/50, 2N2836-2869-3611.
2N301A	AD132-138/50, OC22, 2N2870-3611.
2N302	AC128-131, 2N2431-4106.
2N303	ASY27, 2N2431, 2SB34.
2N307	AC122-128-162, 2N4106.
2N307A	OC26-30, OD603, 2N2836, 2SB41.
2N309	AF126-131-137.
2N310	AF136.
2N311	AC128-131, 2N4106, 2SA156.
2N312	AC128-131, 2N4106.
2N313	2N292.
2N314	2N293.
2N317	2N282-592.
2N318	AF126-135-136.
2N319	AF117-151-153-180, 2N187A-3431.
2N320	AC151-180-190, 2N188A-2429.
2N321	AC151V-180-190, 2N214A-2429, 2SB226.
2N322	AC180, ASY26, 2N190-1130-2431.
2N323	AC180, ASY26, 2N191-2431.
2N324	AC118-180, 2N192-2431-4106.
2N325	AD139-148-152, 2N2836-3611.
2N327	BC177-204-210-212-216-261, BCY27, GFT31/15, SFT124, MPS3703.
2N327A/B	MP33703.
2N328	GFT31/15, MPS6522, SFT124.
2N328A/B	MPS6522.
2N329	GFT31/15, MPS3703, SFT124.
2N329A	BCY29-32, MPS3703.
2N329B	MPS3703.
2N330	GFT31/15, MPS3702, SFT124.
2N330A	MPS3702.
2N331	2N1502-4106.
2N332	BC140-6, MPSH34.
2N332A	BC140-6, BFY10, MPSH34.
2N333	ASY14, BC140-140/6, BSY44, MPSH34, 2N1613.
2N333A	MPSH34.
2N334	ASY27, BC140-140/6, BSY44, MPSH34, 2N1613.
2N334A/B	MPSH34.
2N335	ASY14, BC140-140/0, BSY44, MPSH34, 2N1613.
2N035A/B	MPSH34.
2N336	ASY14, BC140-140/6, BSY44, MPS3710, 2N1613.
2N336A	BC140-140/10-140/16, BSY11-44, MPS3710, 2N1613.
2N337/A	MPS3710.

2N338	ASY27, BC107A, BFY11, MPS3709.
2N338A	BSY11, MPS3709.
2N339	ASY10; MPS3709.
2N340	MPSOA6.
2N340A	BSX46.
2N341	BF114-178, BFY43, MPSOA6.
2N342	BSY45, MPSOA5, 2N1893.
2N342A/B	MPSOA6.
2N343	BCY59, MPSOA5.
2N343A/B	MPSOA5.
2N344	AF131-153-195, 2N404.
2N345	AF131-153-195, 2N274-404.
2N346	2N404.
2N350	AD138-150, OC30, TI3027, 2SB41.
2N351	AD138-150, TI3027, 2N2869.
2N352	AD131-138/50, 2N1536-2836.
2N353	2N1536-2836.
2N354	MPS6522.
2N356	ASY73, 2N1302.
2N357	2N1302.
2N358	ASY27-28-48-75, 2N1304.
2N359	AC117R, 2N652.
2N360	AC117-128-153, HJ17D, 2N369-1192.
2N361	AC117-128-131-153-180, 2N363-1192, 2SB226.
2N362	AC131-151, 2N1192-2429.
2N363	AC122-128-131-151, 2N1191-1192-2492, 2SB227.
2N364	2N2430.
2N365	2N2430.
2N366	AC127, 2N2430.
2N367	AC153, 2N1191-2492.
2N368	AC122-125-151-153, 2N1191-2429.
2N369	2N1191-2429-4106.
2N370	AF125-127-133-135-137-197, 2N3324.
2N371	AF127-132-136-196, 2N3324.
2N372	AF127-133-136-197, 2N3324.
2N373	AF117-127-131-133-136-197, 2N2188.
2N374	AF125-127-133-135-137-197, 2N1639-3325.
2N375	AC105-128, AD131-138/50, 2N4106.
2N376	AD150, 2N2836.
2N377/A	AC153V, ASY29, 2N1302.
2N378	AD133-138, TI3027, 2N4106.
2N379	AD138, ASZ12, AUY22, TI3027-3029.
2N380	AD138, AUY22, TI3027-3030.
2N381	AC117R-153, 2N1305-4106.
2N382	AC125.
2N383	AC125-131, 2N4106.
2N384	AF106-118-137, BF340, GMO760, 2N2189-3325-4034.
2N385	AC153V, ASY29, 2SD11.
2N386	AC125, AD131-138-149, MP869, 2N651A-1531.
2N387	AC125, 2SB247.
2N388	AC127, 2N375.
2N389	BDY39, BUY13, 2N3055-3445.
2N391	AC125, 2N651A.
2N392	AD138/50, AUY21-21IV-28, TI3027, 2N1550.

2N393	2N967.
2N394	AC152V, OC76-80.
2N395	2N581-1308.
2N396/A	AF137, ASY26-27, OC45, 2N1305.
2N397	ASY27, 2N1307.
2N399	AD150, TI3027, 2N2836.
2N400	AD131, TI3027, 2N2836.
2N401	AD131, AUY19, SFT232, 2N2836-3611.
2N402	AC128-153, 2N1191-1924.
2N403	AC117-153, 2N1191.
2N404	ASY27, MPS404.
2N404A	ASY27, MPS404A.
2N405	AC122R-125-128-151IV, 2N322-406-4106, 2SB32.
2N406	AC122G-125-151-151V, 2N322-323-2429, 2S44.
2N407	AC117R-126-131-151VI, 2N324-408-2431.
2N408	AC126-131-151VI, 2N324-382-2431.
2N409	AF126-132-136-190-196, 2N1638, 2S31.
2N410	AF132-136-196, 2N1638.
2N411	AF133-137-197, 2N1639.
2N412	AF127-133-137-197, 2N1639.
2N413	AF126-137-190, HJ56, SET316.
2N413A	ASY26, 2N218.
2N414	AF126-137-188, ASY26, SET316, 2N218.
2N414A	AF126-188, 2N218.
2N414B	2N1307.
2N415	AF126-127-133-137-187-197, SET316, 2N271-274-374.
2N415A	2N374-1307.
2N416	AF126-137-188, SET316, 2N247-1309.
2N417	AF126-137-188, SET316, 2N247-1309.
2N418	2N1100-1537.
2N419	AC125-128, 2SB248A.
2N420	ASZ18, AUY21II-28, 2N1535.
2N420A	ASZ18, 2N1537.
2N422	AC125-128, 2N651-2429.
2N424	2N3446.
2N425	ASY22-27, 2N1305.
2N426	AF136.
2N427	AFY15, 2N1307.
2N428	AFY15.
2N438A	2N1302-1304.
2N439A	2N357-1306.
2N440	ASY28.
2N440A	2N358-1306.
2N442	ADZ11-12.
2N443	ASZ16.
2N444	ASY28.
2N445/A	ASY28.
2N446	2N357.
2N450	AFY15, 2N1308, 2SA206.
2N456	AD130-138.
2N456A	ASZ16.
2N457	AD131-138/50, ASZ17, MP251.
2N458	AD132-138/50, AUY21II-22II-28, ASZ18, 2N561.
2N459	2N378.

2N460	AC128-131, 2N2431-4106.
2N462	OC73, 2N331.
2N463	AD156, 2N1551, 2SB107A.
2N464	AC122-128-162. 2N2431-4270.
2N466	AC128-151VI, 2N2431, 2SB220.
2N467	AC128-171, 2N2431.
2N470	BFY49, BSX45/6, MPS2711, 2N706.
2N471	BFY10-49, BSX45/6, MPS3707, 2N706.
2N471A	BFY10-49, MPS3707, 2N706.
2N472	BFY10-49, BSX45/6, MPS3707, 2N706.
2N472A	BFY10, MPS3704.
2N473	BFY49, BSX45/6, MPS3704, 2N706.
2N474/A	BFY11-69, MPS3707, 2N706.
2N475	BFY11, BSX45/6, MPS3707, 2N706.
2N475A	MPS3704.
2N476	BCY65, MPS3704, 2N706.
2N477	BCY65, BSY11, MPS2711, 2N706.
2N478	BCY65, BSY11, MPS3704, 2N706.
2N479	BCY65, BSY11, MPS2711, 2N706.
2N479A	MPS3704.
2N480	BCY65, BSY11, MPS3704, 2N706.
2N480A	MPS3704.
2N481	AF126-132-136-185-190-197, AFY15.
2N482	AF126-132-136-185-190-197, AFY15, 2N2189.
2N483	AF126-132-136-185-190-197, AFY15, 2N373-2189.
2N484	AF126-132-136-185-197, AFY15, 2N373-2190.
2N485	AF126-132-136-185-190-197, AFY15, 2N374-2190.
2N486	AF126-132-136-197.
2N495	BC177-211, BCY28, BCZ11, MPS6580, 2N354-2945.
2N496	MPS3640.
2N499	AF106-126-132-136-137-197, 2N371-2188.
2N500	2N2189-3323.
2N501	AF102-106-137, 2N2189.
2N502	AF106, GMO760, 2N2189.
2N503	ASZ21, 2N2189-3284.
2N504	AF124-134, 2N2189.
2N506	AC125-151.
2N508	AC117-128-153-180, 2G508.
2N509	2N1195.
2N511	2N456A.
2N512	2N456A-1558.
2N513	2N456A-1163.
2N514	ADY26, 2N514-1100-1163.
2N515	ASY26, 2N1304.
2N516	ASY26-73, 2N1304.
2N517	ASY26-73, 2N1306.
2N519	AC117R-153, 2N404-578.
2N520	ASY26, 2N404-578.
2N521	2N579-1377.
2N522	2N522A-1309.
2N523	2N1305-1377.
2N524	AC128-131/30-131/50-153, ASY48IV.
2N525	AC128-131/30-152, ASY26-48IV, 2SB225.
2N526	AC131/30, ASY26-48IV-80.

2N527	AC131/30-131/50, ASY48IV-80, 2N526-1307.
2N527A	2N1307.
2N529	2N1303.
2N530	2N1303.
2N531/2	2N1305.
2N533	2N1305.
2N535	AC122-122G-126-151VI, AF181, ASY48IV, OC304 2N404-1192-1193.
2N535A	AC122G-126-151VI, OC304, 2N1193.
2N535B	AC122W-126-151, OC305.
2N536	2N204-404-1193.
2N537	AF139.
2N538	AUY21II-28, TI3027, 2N2140.
2N539	TI3027, 2N2145.
2N540	AD131-138/50, TI3027, 2N1551.
2N541	BCY59, BSY11-44, MPS3393.
2N541A	MPS3393.
2N542/A	MPS3707.
2N543	MPS3704.
2N543A	MPSA10.
2N544	AF138-185, MPSA10.
2N549	2N1893-2243A.
2N550	2N1893-2243A.
2N551	2N1893-2243A.
2N552	2N1893-2243A.
2N554	AD149IV, TI301-3027.
2N555	AD149IV, TI3027.
2N556	ASY75, 2N1302.
2N557	ASY75, 2N1304.
2N558	ASY75, 2N1306.
2N559	2N645.
2N561	AD131-149, 2N456A, 2SB249.
2N563	AC128-153, 2N650-1924-2431, 2SB103.
2N564	AC128, 2N650-2431.
2N565	AC128, 2N651-2431.
2N566	AC128, 2N651-2431.
2N567	AC128, 2N651-2431.
2N568	AC117-128, 2N651-2431.
2N569	AC128, 2N1193-2431.
2N570	AC117R-153, 2N1192-2431.
2N571	AC117-R153, 2N1193-2431.
2N572	AC117-R153, 2N1193-2431.
2N573	AC125-131-151-184.
2N574	AUY20VI-22, 2N456A-1022A.
2N575	AUY20VI-22, 2N1554.
2N576	ASY26-73, 2N1304-1306.
2N578	2N404.
2N579	2N404.
2N580	2N1307-1309.
2N583	AГY12.
2N585	ASY28.
2N586	AC125-128-131-151-184, 2N1191, 2SB225.
2N588	2N3324.
2N589	ASZ21, 2N1532-3324.

2N591	AC117-122/30-153-153VI, 2N1192-2429.
2N592	AC153.
2N593	AC153, 2N604-608.
2N597	AC128-131, 2N3427.
2N598	2N1997-1998.
2N599	2N1997.
2N600	2N1997-1998.
2N601	2N1997-1999.
2N602	AC128, ASY24, 2N2635.
2N603	AC128-131-152, ASY24, 2N644-2635, 2SB65.
2N604	AC128-131-152, ASY24, OC76, 2N645-2635, 2SB65.
2N605	AC128-131-152, ASY24, OC307, 2N608-2635, 2SB65.
2N606	AC128-131-152, ASY24, OC76, 2N603-2635, 2SB65.
2N607	AC128-131-152, ASY24 OC76, 2N605-2635, 2SB65.
2N608	AC128-131-152, ASY24, OC72, 2N605-2635, 2SB65.
2N609	AC117R-153, 2N404.
2N610	AC128, 2N404.
2N611	AC117-153, OC72, 2N217-1309, 2SB222.
2N612	AC122-125-128-151, OC304, 2N1191.
2N613	AC122-131-152, ASY26, 2N217, 2SB224.
2N614	AF137-185, 2N4034.
2N615	AF127-137-185, 2N373-4034.
2N617	AF106-185, AFY15, 2N374.
2N618	AD138/50.
2N619	MPS3704.
2N620	MPS3704.
2N621	MPS3708.
2N622	MPS3704.
2N623	AF134.
2N628	TI3027, 2N561.
2N629	TI3027.
2N630	TI3027, 2N1014.
2N631	AC124, 2N404.
2N632	2N404.
2N633	AC117-131-153, OC72-76, 2N404, 2SB37-220.
2N634A	2N1304-1306.
2N635/6	2N1304.
2N638	AC131-153, OC76, 2N561, 2SB248A.
2N638A/B	2N561.
2N639	2N561.
2N640	AF125-136-196, 2N1637-2188.
2N641	AF136-196, 2N1638.
2N642	AF126-132-136-137, 2SA114.
2N643/4/5	2N1309-2635.
2N647	2N1306.
2N649	2N1308.
2N650	2N1997.
2N651/A	AC117-128-153, 2N1997, 2SB247.
2N652/A	AC117-128-153, ASY26, 2SB222-247.
2N653/4/5	2N1997.
2N656	BSX22-45-46/6, 2N656-4331.
2N658	AC117-153, 2N2000.
2N659	AC117-128-153, AD162, 2N2000.
2N660	AC117-128-153, AD162, 2N643-2000.

2N661	AD162, 2N2000.
2N662	AC117-128-153, AD162, OC76, 2N2000.
2N665	TI3027.
2N669	TI3027.
2N670	2N1038.
2N671/2/3/4	2N1038.
2N677	2N456A.
2N678	AD133, ADZ11, 2N456A.
2N679	2N1304.
2N680	AC117-128-153, 2N404-1309, 2SD11.
2N695	ASZ21, 2N2635.
2N696	BSX45-45/6, BSY44, MPS6530, 2N696-697-2218-2218A.
2N696A	MPS6530.
2N697	BSX45-45/10, BSY44-51, MPS6530, 2N2218-2218A.
2N697A	MPS6530.
2N698	BF178, BFY45, BSW67, BSX46/10-47/6, 2N3498-4410.
2N699	BF114-178, BFY41-45, BSW67, BSX47/6-10, 2N4410.
2N699A/B	2N4410.
2N700	AF139, 2N2415-5043.
2N702	MPS6512.
2N703	2N4124.
2N705	AF106, AFY11-19, ASZ21, GMO760, 2N964.
2N706	BSX19, BSY62/A, MPS706, 2N706A-964.
2N706A	BSX19, MPS706A.
2N706B	MPS706A.
2N706C	MPS2369.
2N707/A	MPS3826, 2N2483-2484.
2N708	BSX19, MPS834.
2N708A	MPS834.
2N709	AF66, BSX19, BSY18, 2N2369A-3303.
2N709A	2N3303.
2N710	2N964.
2N711	AC125, AF106, AFY11, 2N964.
2N715/6	MPS6530, 2N4875.
2N717	BSX45/6, MPS6530, 2N2221A-2222.
2N718	BSX45/6-10, MPS6531, 2N2221A-2222.
2N718A	BSX45-6/10, MPSH04, 2N2221A.
2N719/A	BSX47-6, 2N2222-4410.
2N720/A	BF178, BSX47-6/10, 2N2222-3498-4410.
2N721/A	MPS3703, 2N2907.
2N722/A	MPS3703, 2N2907.
2N725	AF121-201-202, 2N2635.
2N726/7	BC178VI, MPS6516.
2N728	BSY62-63, 2N4264.
2N729	2N4123.
2N730/A	BSX45-6/10. MPS6530.
2N731	MPSA10, 2N2221A.
2N733	2N2221A.
2N734	BFY80, BSX46, MPS6530.
2N734A	MPSA05.
2N735	BFY80, BSX46-70, MPS6530, 2N1091-2221.

2N735A	MPS6530.
2N736/A	BFY80, BSX46-71, MPS6531, 2N1091-2222A.
2N738/A	2N4410.
2N739/A	2N2221A-4410.
2N740	BC141, 2N2222A-4410.
2N740A	BFY45, 2N4410.
2N741	AF121S-202/S, 2N2996-5043.
2N742/A	AF202, MPS6530, 2N2217-2219.
2N743	BSX19, BSY17-21, 2N4265.
2N743A	2N4123.
2N744	BF168, BSX20, BSY18-21, 2N4265.
2N744A	2N4123.
2N745	MPS3708.
2N746	MPS3709.
2N747/8	2N4124.
2N749	BCY59, BSY44, MPS3709, 2N697.
2N750	BCY59, BSY11-44.
2N751	MPS2711, 2N697.
2N752	BFY44, MPS6530, 2N736.
2N754/5	MPS6530, 2N1893-2243A.
2N756	BSY10, MPSA10, 2N734.
2N756A	MPSA05.
2N757	BSY10, MPSA10.
2N757A	MPSOA5.
2N758	MPSA10, 2N734.
2N758A/B	MPSA05.
2N759	MPSA10.
2N759A/B	MPSA05.
2N760	BCY59VII-59IX, MPSA10.
2N760A/B	BCY65VIII, MPSA05, 2N2483.
2N761/2	MPS3705.
2N770/1	MPS6568.
2N772	MPSH10.
2N773/4/5	MPS6568.
2N776/7/8	MPS6568, 2N734.
2N779	2N964.
2N780	BC107A, BF115, MPS6530.
2N781/2	2N2635.
2N783/4	2N2369A-4264.
2N784A	MPSH34.
2N789	MPS3705.
2N790	MPS3705, 2N333.
2N791	MPS3705, 2N334.
2N792	MPS3705, 2N335.
2N793	MPS3705, 2N336.
2N794/5/6	2N2635.
2N800 to 810	2N404.
2N812/3/4	2N404.
2N825/6	2N404.
2N834	BCY56, BFY66, BSY63, MPS834, 2N3014.
2N834A	2N426A.
2N835	BSY62, MPS835, 2N3014.
2N838	2N2635.
2N839	BF177, BSY44, MPS6531, 2N929-930.

2N840/1	MPS6530, 2N929-930.
2N842	MPS6530, 2N4252-4253.
2N843	MPS6531, 2N4252-4253.
2N844	MPS6565, 2N1893-2243A.
2N845	2N1893-2243A-4410.
2N846/A/B	2N964.
2N847	MPS2711.
2N848	MPS3394.
2N849	BSX19-48, MPS835.
2N850	BSX20-48, MPS834.
2N852	MPS384.
2N858/9	MPS404A, 2N2945.
2N860/1	MPS3638, 2N2944-2945.
2N862/3	MPS3640, 2N2944-2945.
2N864/A	MPS3639, 2N2904-2905.
2N865/A	MPS3640, 2N2904-2905.
2N866/7	2N4124.
2N869	MPS3640, 2N2904-2905.
2N869A	MPS3640, 2N289-3576.
2N870	BSX46/6-10, 2N2222-4410.
2N871	BSX46/10, 2N2222-4410.
2N902	MPS3708, 2N332.
2N903	MPS3705, 2N333.
2N904	MPS3705, 2N334.
2N905	MPS3705, 2N335.
2N906	MPS3705, 2N336.
2N907	MPS3705, 2N337-3015.
2N908	MPS3705, 2N338-3015.
2N909	MPS6531, 2N2192-2243A.
2N910	BFY65-80, BSX46/10. MPSA06, 2N1973-2222.
2N911	BFY80, BSX46-46/10, MPSA06, 2N1974-2222.
2N912	BSX46/6, MPSA06, 2N2222.
2N914	BCY59, BSX20-88, BSY63, MPSH32.
2N914A	MPSH32.
2N915	BCY65, BF115-194-225, BSW63, MPS3826, 2N2221A.
2N916	BCY56, BFY27, BSY63, MPS6512.
2N916A	MPS6512.
2N917	MPS3563, 2N918.
2N917A	MPS3563.
2N918	BFX62, BSX19, MPS918, 2N3600.
2N919to922	MPS834.
2N923/4	MPS3702.
2N925/6	MPS3703.
2N927/8	MPSA55, 2N2604-2605.
2N929	BC107-107A, BFX92A, MPS6512-6513-2N930.
2N929A	MPS6514.
2N930	BC107B, BCY59, BFX93A, MPS6514.
2N930A/B	2N5210.
2N934	2N2635.
2N935	MPSA70, 2N327A-2945.
2N936	MPS3703, 2N328A-2945.
2N937	MPS3703, 2N329A-2945.
2N938/9/40	MPS3702, 2N2945.
2N941/2	2N2945-5221.

2N943/4	MPS6533, 2N2945.
2N945	2N2945-5086.
2N946	MPSH54, 2N2945.
2N947	BSY62-63, MPS834, 2N834.
2N955	2N797.
2N956	BC140C-141/10, BSY83-93, MPS6531, 2N2222/A
2N957	MPS6512, 2N2484-2501.
2N958/9	2N706-3303.
2N960/1/2/3	2N964.
2N964/A	ASY27.
2N965 to 975	2N964.
2N976	2N961-964.
2N977	2N964-985.
2N978	MPS6579, 2N721-2907.
2N979/80	2N2635.
2N981	MPSA06.
2N982/3/4	2N964-985.
2N986	MPS6569, 2N964.
2N987	AF106, 2N2635.
2N988/9	MPS6530, 2N706.
2N990	AF130-134-178-194.
2N991	AF131-135-136-185-195, 2N2635.
2N992	AF132-136-185-196, 2N2635.
2N993	AF133-137-185-197, 2N2635.
2N995	BC179-179VI-206-263, BSW19, MPS6580.
2N995A	MPS3640.
2N996	2N2906-2907-5221.
2N997	MPSA20.
2N998/9	2N997.
2N1005	MPS2711.
2N1006	MPS2711.
2N1007	ADY27.
2N1008	AC125-131-151-184.
2N1008A	AC125-131/30-151.
2N1008B	AC125-151, ACY24.
2N1009	AC117-128-153, 2SA219.
2N1010	2N1302.
2N1011	ADY26, AUY21II, TI3027.
2N1012	2N388A-1306.
2N1014	AC128-183, ASZ18, 2N456A-1021.
2N1016	2N3713.
2N1017	2N404.
2N1018	2N404-582.
2N1021	2N456A-1014.
2N1022	2N456A-1100.
2N1023	AF106, 2N2635.
2N1024	BC160, MPS6522, 2N1132-2944-2945.
2N1025	BC160, MPS3702, 2N1132-2944-2945.
2N1026	BC160, BCY11-12, MPS3702, 2N1132-2944-2945.
2N1027	BC178-178VI-205, MPS6522, 2N2944-2945.
2N1028	MPS6522, 2N2944-2945.
2N1029/A-B	TI3027-3031.
2N1029C	2N456A-3146.
2N1030/A	2N456A-514A/B.

2N1031/A	AD133, ADZ11, TI3027.
2N1032/A	2N456A-514A/B.
2N1034/5/6/7	MPS3703, 2N2944-2945.
2N1038	AD149-149IV, 2N2552.
2N1039	AD138/50, ASZ17, AUY21III, OC26-30, 2N1038-1041, 2SB181.
2N1040	AD149-162, OC26-30, OD603, 2SB181, 2N1038-1041.
2N1041	AD138/50, ASZ15, AUY21IV, 2N1038.
2N1043	AD138/50, ASZ15, OC26-30, 2N1038-1040, 2SB181.
2N1044	AD138/50, 2N1038-1043, 2SB181.
2N1045	2N1014-1038.
2N1046/A-B	AC128-153, ACY24, 2N1056-1907, 2SB224.
2N1051	MPSH20, 2N2217-2219.
2N1052	2N5058.
2N1054	MPS401, 2N5059.
2N1055	2N4410.
2N1056	ACY24.
2N1057	AC125-128-131/30-151.
2N1058	ADY26.
2N1060	2N2217-2219-4123.
2N1065	ACY24, ASY48-81.
2N1069/70	2N4913.
2N1072	2N3766.
	AD133-138, ADZ11.
2N1074/5/6/7	MPS3704, 2N328A.
2N1081	2N3724-3725-4123.
2N1082	MPS6568, 2N2217-2219.
2N1086/A	2N1308.
2N1087	2N1308.
2N1090/1	2N1304-1605.
2N1092	2N4264.
2N1093	AF101-106-126-127-178-190, 2N1305, 2SA206.
2N1094	2N1308.
2N1097	AC128-131, 2N1414.
2N1098	AC128, 2N1414.
2N1100	AD132V, 2N3771-3772.
2N1101	AC179. 2N647-1302.
2N1102	AC179. 2N647-1306.
2N1103	MPS3708.
2N1104	MPS3709.
2N1105	MPSA05.
2N1106	2N4410.
2N1107	AF126-132-136-196.
2N1108	AF125-127-133-137-181-197, 2N2188.
2N1109	2N2188.
2N1110	AF125-127-133-137-197, 2N2188.
2N1111/A-B	AF127-133-197, 2N2188, 2SA156.
2N1116	BFY34, BSX22-45, MPSA05, 2N2243A-3252.
2N1117	MPSA05, 2N2103-2243A.
2N1118/A	MPS6519, 2N2604-2605.
2N1119	MPSL08, 2N2604-2605.
2N1120	AD133, TI3027-3031.
2N1121	2N1306.
2N1122/A	AF126-132-137, 2N961-964, 2SA155.

2N1123	2N1997-3427.
2N1124	2N651-1377.
2N1125	2N651-2000.
2N1126	2N651.
2N1128	2N1377.
2N1129	AC117-128-153, 2N1377-1379, 2SB222-223.
2N1130	AC117-128-153, 2N1377, 2SB221-222.
2N1131	BSV16-6, 2N1132-2904-2905-4402.
2N1131A	2N3905.
2N1132	BCX10, BCY18, BCZ11-12, BSV16-6, 2N2904-2905-3905.
2N1132A	2N3905.
2N1132B	MPSH55.
2N1135/A	MPSL08.
2N1136	AD133-138, ADZ11, ASZ18, AUY21, MPS249, TI3027.
2N1136A	AUY22-28.
2N1136B	AUY21.
2N1137	AC117-153, AD138, ASZ16-18, AUY21, MP438, 2N456A, 2SB248.
2N1138	AD130V-138-149, MP280.
2N1139	MPS706.
2N1141A	AFY18.
2N1142	AFY12.
2N1142A	AFY18.
2N1143	AFY12, AFZ12, ASY76.
2N1143A	AFY18.
2N1144	AC128-153, 2N321-1303.
2N1145	AC131-152-153, OC76, 2N1303-1414, 2SB37.
2N1146	AD133-138, ADZ11, 2N456A.
2N1146A	AD138-149-153, ADZ11-12, AUY21-21IV, 2N456A.
2N1146B	AD138-153, AUY22-22IV.
2N1146C	AD131-138-138/50, ADZ12, 2N456A.
2N1149/50	MPS3708.
2N1151/2	MPS3709.
2N1153	MPS3710.
2N1154	MPS3705.
2N1155	MPSA06.
2N1156	MPSL01.
2N1157	MP501.
2N1157A	MP502.
2N1158	2N1143.
2N1158A	2N1142.
2N1159	ASZ25, AUY22III-28, 2N456A-3146.
2N1160	ASZ16, AUY22, 2N456A-3146.
2N1168	ASZ18, AUY19III-21, TI3027, 2N3614.
2N1172	AD139-148-156, OC22-30, TI3027-3028, 2SB240.
2N1174	BC177, BCY28, 2N404.
2N1175	ASY26.
2N1176	2N1038.
2N1177	AF125-178, 40242, 2N2188.
2N1178	AF125-178, 40244, 2N2188-2955, 2SA118.
2N1179	AF115-118-125-136-178, 40243, 2N2188-2956, 2SA118.
2N1180	AF115-118-125-136-178, AFZ12, 2N2188-2956, 2SA118.

2N1183	AC136, AD136-148-152-160, AUY18, 2N1038-2140.
2N1183A	AD148-152, AUY18, 2N1038.
2N1184/A	AD148-152, 2N1038-2564.
2N1184B	2N1038-2565.
2N1185/6	ASY26, 2N1375-1377.
2N1187/8	2N1376-1377.
2N1189/90	2N1377.
2N1190S	BSW65.
2N1191	AC128, 2N404.
2N1192	2N404.
2N1193	AC117-128-153, OC318, 2N404, 2SB227.
2N1195	AF102-106-178, AFY12, 2N404
2N1196/7	MPSH54.
2N1199	2N3303.
2N1199A	MPS834.
2N1200/1	MPSH02, 2N4252.
2N1206	MPS6544, 2N5059.
2N1207	MPSL01, 2N5059.
2N1208	2N3487.
2N1209	2N1724.
2N1210/1	2N1722.
2N1212	2N1724.
2N1217	2N1308.
2N1219	BC160, MPS3702, 2N1132.
2N1220	BC160, BCY12, MPS3702, 2N1132-1219-2904-2905.
2N1221/2	MPS3702, 2N2904-2905.
2N1223	MPS3703, 2N2904-2905.
2N1224	AF126-127-133-137-197.
2N1225	AF125-135.
2N1226	AF114-118-124-127-133-137-197, 2N2635.
2N1227	AD149, NKT404, TI3027, 2N3611.
2N1228/9	MPSL08.
2N1230/1	MPS404A.
2N1232/3	2N4402.
2N1234	MPSL51.
2N1238/9	MPS3640.
2N1240/1	2N3905.
2N1242/3	2N5086.
2N1244	MPSL51.
2N1245	AD162, SFT232, 2N404.
2N1246	2N404.
2N1247/8/9	MPS2711.
2N1250	2N3713.
2N1251	AC127, 2N1304.
2N1252	AC127-179, BFY50, BSX45, MPS3646, 2N1304-2218.
2N1252A	2N3903.
2N1253	BFY51, BSX45, MPS3646, 2N2218.
2N1253A	2N3903.
2N1254/5	2N722-2007-4125.
2N1256	2N722-2907-4402.
2N1257	BC177-261, MPS6534, 2N722-2907-4402.
2N1258/9	2N722-2907-3905.
2N1261	AC105, 2N1262-1531.
2N1262	AC105-117-128-131-153, 2SB252, 2N1351.

2N1263	AUY22-28, 2N458-3617.
2N1264	AF127-137, 2N1191.
2N1265	2N404-1192.
2N1266	AC122-125-151, AF122-128-185, 2N404-1189-1191.
2N1267 to 72	2N4252-4253-5219.
2N1273	AC128, 2N404.
2N1274	AC117-122/30-125-151-153, ASY26, OC72, 2N404, 2SB222.
2N1275	MPS456.
2N1276 to 9	MPS3708, 2N2501.
2N1280	2N1305.
2N1281/2	2N1307.
2N1284	2N1305.
2N1285	2N2188.
2N1287	AC137, 2N651-1303.
2N1291	AC117-128-153, OC318, TI3027, 2N1529, 2SB248.
2N1293	AC117-128-153, OC74, TI3027, 2N1531, 2SB248A.
2N1295	AC117-128-153, OC74, 2N456-456A-1532, 2SB249.
2N1297	2N456A-3146.
2N1298	2N1302.
2N1299	2N1306.
2N1300/1	ASY27, ASZ20, 2N2635.
2N1302	ASY26-28-73, BSX19-20.
2N1303	BCY70-72, 2N1192.
2N1304	ASY26-28-29, BSX19-20, 2N1192.
2N1305	ASY26-27, BCY70-72.
2N1306	ASY26-29, BSX19-20.
2N1307	ASY26-27, BCY70-72.
2N1308	ASY26-29, BSX19-20.
2N1309	ASY26-27, BCY70-72.
2N1310	MP2373-1.
2N1311	MP2373-3.
2N1313	MP2373-5.
2N1314/5	2N3611.
2N1324	AC117-128-153, 2SB249.
2N1326	2N1038.
2N1328	AC117-128-153, 2N1038, 2SB248.
2N1331	AC117-128-153, OC74, 2SB249.
2N1335/6/7	2N5550.
2N1338	BFY13, BSX45-46, BSY46-85, MPSH05, 2N2193.
2N1339	2N4410.
2N1340	BF178, BFY45, BSW66, BSY45, 2N4410.
2N1341	2N4410.
2N1342	BF178, BFY45, BSW66, BSY45, 2N4410.
2N1343 to 51	2N404.
2N1353 to 57	2N404.
2N1358	AC128, AD138/50, ADZ12, ASZ16, AUY21-29, 2SB252.
2N1359	2SB249.
2N1360	TI3027.
2N1362	AUY34, 2N456-456A-3146.
2N1363/4/5	2N456A-3146.
2N1366	2N1302.
2N1367	2N1304.

2N1370	ASY26, 2N1309-1377.
2N1371	ACY19, 2N1377.
2N1372	2N1307-1377.
2N1373	ACY19, 2N1377.
2N1374	ACY19, ASY26, 2N1377.
2N1375	ACY19, 2N1377.
2N1376	ACY19, ASY26.
2N1377	ACY19.
2N1378	2N1309-1377.
2N1379/80	ASY26, 2N1193-1309-1377.
2N1381	2N1309.
2N1382/3	ACY19, 2N1377.
2N1384	2N2635.
2N1385	AF139, ASZ21.
2N1386 to 90	MPS6530, 2N2218.
2N1392	AC125-151.
2N1394	AC122-125-151-162, OC70, 2N1392, 2SB111.
2N1395	2N2188.
2N1396	2N2191-3323.
2N1397	AF118, ASZ20, 2N2191.
2N1398	2N2996.
2N1399	AF106, 2N2996.
2N1400	AF124-131-134, 2N2996.
2N1401/2	AF124-134, 2N2996.
2N1403	2N2996.
2N1404	AF124-134, 2N1303-1307.
2N1404A	2N1303.
2N1405/6/7	2N2996.
2N1408	2N2000.
2N1409/A	2N4124.
2N1410	2N2124.
2N1410A	BSX46, 2N4124.
2N1411	AF134, 2N962-2188.
2N1412	2N1100.
2N1416	AC117-128-153, ASY26, 2N1193.
2N1417	MPS2711.
2N1418	MPS3710.
2N1419	2N1164.
2N1420	BSX46-10/16, BSY52, MPS6531, 2N2222-2243/A.
2N1420A	MPS6531, 2N2243A.
2N1425	AF126-134-137, 2N2188.
2N1426	AF124-126-134-136, 2N2188.
2N1427	AF124-134, 2N962-2635.
2N1428	MPS3639, 2N1132-2905.
2N1429	BC173-179-202R-238-239-263, MPS3639, 2N1132-2905.
2N1431	2N1302.
2N1432	2N2189.
2N1433	AC117-128-153, 2SB252.
2N1434	AD139-148.
2N1435	AD139-148, OC30, OD603, 2N1434, 2SB259.
2N1437	AC128, 2N456A-3146.
2N1438	2N456A-3146, 2SB252A.
2N1439 to 43	MPS3703, 2N2945-2946.

2N1444	2N3252-3903.
2N1446 to 49	2N1373-1377.
2N1450	2N1143.
2N1451/2	2N1375-1377.
2N1465/6	2N456A-3146.
2N1469	2N2944-2945.
2N1471	2N1309.
2N1472	2N2537-3015-4124.
2N1473	2N2000.
2N1474	BSY16, MPSA55, 2N2944-2945.
2N1474A	BCY11-12, MPSA55, 2N2944-2945.
2N1475	BCY11, MPSA55, 2N2944-2945.
2N1476	MPSL51, 2N2944-2945.
2N1477	2N2944-2945-5401.
2N1478	2N1307-3427.
2N1479	BFX34, 2N2987-4237.
2N1480	BFX34, 2N2988-4238.
2N1481	BFX34, 2N2989-4239.
2N1482	BFX34, 2N2990-4238.
2N1483 to 86	2N3766.
2N1487	BD130, BDY38, 2N3055-4913.
2N1488	BD130, BDY20, 2N3055-4914.
2N1489	BD130, BDY38, 2N3055-4913.
2N1490	BD130, BDY20, 2N960-3055-4914.
2N1491	2N2218.
2N1493	BSY62, 2N2218.
2N1499	2N964.
2N1499A	AF106-190, 2N960-964.
2N1499B	2N964.
2N1500	2N964.
2N1501	AC118-128, 2N2144, 2SB248A/B.
2N1502	AC117-128-153, AD138-149, TI3027, 2N2143, 2SB248.
2N1504	2N456A-3146.
2N1505	BFY12, BSX45-45/6, MPS6530, 2N2217-2218-2219.
2N1506	BFY12, BSX45-45/6, MPS6565, 2N2217-2218-2219.
2N1506A	2N4409.
2N1507	BFX85, BFY68A, BSX16-45/10, BSY44, MPS6530, 2N2219-2243A.
2N1508	2N4410.
2N1509	MPS6530.
2N1515	AF132-137-185-196.
2N1516	AF125-126-131-135-137-185-195.
2N1517	AF124-130-134-137-185-194, 2N2495.
2N1518	MP943.
2N1519	MP943A.
2N1520	MP944.
2N1521	MP944A.
2N1524	AF127-133-137-197.
2N1525	AF127-133-137-197-200, TIS37, 2N3325.
2N1526/7	AF127-133-138-197, TIS37, 2N3325.
2N1528	BSW51, BSY34, 2N2218.
2N1529/A	AUY19III.
2N1530/A	AUY19III.

2N1531/A	AUY20III.
2N1532/A	AUY34III.
2N1533	AUY34III.
2N1534/A	AUY19IV.
2N1535	AD133, ADZ11, AUY19IV.
2N1535A	AUY29V.
2N1536/A	AUY20IV.
2N1537/A	AUY34IV.
2N1538	AUY34IV.
2N1539/A	AUY19IV.
2N1540/A	AUY19IV.
2N1541/A	AUY20V.
2N1542/A	AUY34V.
2N1543	AUY34V.
2N1544/A	ASZ16, AUY19V.
2N1545/A	ASZ18, AUY19V.
2N1546/A	AUY20V.
2N1547/A	AD131-138/50-149, AUY34V, 2N1532.
2N1548	AUY34V.
2N1549/A	AUY29III, TI3027.
2N1550/A	AUY29III, TI3028.
2N1552/A	2N1021A.
2N1553/A	AUY29IV, TI3027.
2N1554/A	AUY29IV, TI3028.
2N1556/A	2N1021A-1100.
2N1557/A	AUY29V, 2N456A-514.
2N1558/A	AUY29V, 2N514A.
2N1559/A	2N514B.
2N1560/A	2N1021A.
2N1564	BFY67A-80, BSX46, MPS6530, 2SC154, 2N2218.
2N1565	BFY50-67A-80, BSW51, BSX46, MPS6530, 2SC154, 2N2218.
2N1566	BFX85, BFY68A-80, BSW52, BSX46, MPS6530, 2SC154, 2N2219.
2N1566A	BC141/10, MPS6530, 2N2219.
2N1572	BF177, BFY14B, BSX47/6, 2SC857, 2N4410.
2N1573	BF178, BFY14B/C, BSX47/6, 2SC857, 2N4410.
2N1574	BFY14C/D, BSX47/10, 2N1893-4410.
2N1586	MPS2711.
2N1587	MPS3706.
2N1588	MPSA10.
2N1589	MPS2711.
2N1590	MPS3703.
2N1591	MPSA10.
2N1592	MPS2711.
2N1593	MPS3706.
2N1594	MPSA10.
2N1605/A	ASY29, 2N1304.
2N1606/7/8	MPS3640.
2N1613	BFY41, BSX45/6, BSY53, MPS6530, 2SC708, 2N2102.
2N1613A	MPS6530.
2N1613B	BSX47/6, 2N4410.
2N1614	ACY24, ASY48, 2N1924-2000.

2N1615	2N4410.
2N1616	2N1734-3055.
2N1617	BCY28, BDY11-91, BFY38, 2N1211-1724.
2N1618	BDY11-90, BFY34, 2N1724-3488.
2N1620	2N1724-3446.
2N1623	MPS3708.
2N1624	2N1308.
2N1631/2	AF125-131-135-195, 2N2635-3325.
2N1633	AF106-126-132-136-185-196, 2SA230, 2N2635.
2N1634/5/6	AF106-125-126-131-132-135-136-185-195-196, 2SA230, 2N2635.
2N1638	AF130-134-136-194, 2SA235, 2N2635-3325.
2N1639	AF124-127-130-134-135-194, 2SA235, 2N2635-3325.
2N1640/1/2	MPS3638, 2N2945-2946.
2N1643	MPS3702, 2N2944-2945.
2N1644	2N1893-2243A.
2N1646	AF106-201-202, 2SA230.
2N1647	2N2150-4409.
2N1648	2N2151-4410.
2N1649	2N2150-4410.
2N1650	2N2151-4410.
2N1654/5/6	2N2944-2945-4410.
2N1660/1/2	2N1722.
2N1663	MPS2369, 2N2369A-3011.
2N1666	AUY22II/III, TI3027, 2SB341.
2N1667	ASZ16, AUY21/III/IV-23III-28, TI3027, 2SB341.
2N1668	AD130-138-149, ASZ16-17, AUY21IV, TI3027, 2SB341.
2N1669	ASZ16-18, AUY22III, TI3027, 2SB341.
2N1670	2N398.
2N1671A	BC160, 2N1132-1671.
2N1672	2N1302.
2N1673	AF200, 2N404.
2N1674	MPSA20.
2N1676/7	MPS3639, 2N2944-2945.
2N1679	2N4410.
2N1680	2N4400.
2N1682	2N2217-2219-4124.
2N1683	2N2635.
2N1694	2N1302.
2N1700	BFY50, 2N3053.
2N1701	2N3054.
2N1702	2N3055-4913.
2N1704	MPS6530, 2N1893-2243A.
2N1708	2N4265.
2N1708A/B	MPS2369, 2N743.
2N1711	BFY46, BSX45/10, BSY54-71, MPS6530, 2N2219/A.
2N1711A/B	MPS6530.
2N1714	BFY50.
2N1716	BFY51.
2N1718/9/20	2N3766.
2N1721	2N3767.
2N1722	2N3446.
2N1723	BCY17, 2N3448.

2N1724	BLY17, 2N4347.
2N1724A	2N4347.
2N1725	2N3055.
2N1726	2N3323.
2N1727/8	AF134, 2SA235, 2N2996-3324.
2N1729	2N1303.
2N1730	2N1302.
2N1731	2N1303.
2N1732	2N1302.
2N1742	AF106-124-127-137, GMO760, 2N2996.
2N1743/4	2N2996-3284.
2N1745	AF137, 2N2996-3285.
2N1746/7/8	2N2996-3323.
2N1748A	2N2997.
2N1749/50	2N2996.
2N1752	2N2996.
2N1754	AS26-48-77, 2N2996.
2N1755	2N1038-2552.
2N1756	2N1038-2554.
2N1757/8	2N1038-2555.
2N1759	2N1038-2564.
2N1760	AD130-138-149, AUY21III, 2SA341, 2N2143.
2N1761	AUY21II-28, 2SA341, 2N2144.
2N1762	2N2145-2567.
2N1763	MPS2369, 2N3014.
2N1764	2N2369A-3011-4265.
2N1768/9	2N1050-3445.
2N1785 to 90	2N2996.
2N1808	2N1302.
2N1837 to 40	BF194, MPS6530, 2SC260, 2N2218.
2N1841	2N4410.
2N1853/4	2N2635.
2N1864	2N2997-3324.
2N1865	2N2997-3325.
2N1866	2N2997-3323.
2N1867	2N2997-3324.
2N1868	2N2997-3325.
2N1886	BLY17, 2N4914.
2N1889	BC141/10, BFX85, BSS30, BSW65, BSX46/6-10, BSY87, 2N3498-4410.
2N1890	BF109, BFX85, BSS31, BSW66, BSX46-46/16, BSY88, 2N3499-4410.
2N1891	2N1304-1306.
2N1892	2N1303.
2N1893	BFW33, BFY14-45, BSS32, BSX47/10, BSY55, 2SC708, 2N2405-3498-4410.
2N1893A	2N4410.
2N1899	2N4002.
2N1901	2N4002.
2N1905	2N1040-2832.
2N1906	AL100, 2N1907-2832.
2N1917	BC160, MPS404, 2N1132.
2N1918	BC160, BCY12, MPS404, 2N1132-2944-2945.
2N1919	BC212, BCY29, 2N2944-4402.

2N1920	BCY29, BCZ12, 2N2944-2945-4402.
2N1921	BCZ12, 2N2944-2945-3905.
2N1922	BCZ12, MPSH54, 2N2944-2945.
2N1923	MPSH54.
2N1924/5	AC131/30, ASY48-48IV-77, NKT217, 2N1305-1307-2906.
2N1926	AC131/30, ASY77, 2N1307.
2N1936/7	2N3846.
2N1940	TI3027.
2N1941	MPS3705.
2N1943	MPSA05, 2N1893-2243A.
2N1944/5	2N1893-2243A-4124.
2N1946	2N3904.
2N1947	2N4124.
2N1948/9	2N4123.
2N1950	2N4264.
2N1951/2	2N4123.
2N1953	MPS2711.
2N1954	2N651-2000.
2N1955	2N1190-2000.
2N1956	2N651-2000.
2N1957	2N1187-2000.
2N1958/A	2N3903.
2N1959/A	2N3903.
2N1960/1	2N964.
2N1962/3	MPS834, 2N2369A-3011.
2N1964/5	2N2537-3015-3903.
2N1968	2N2997.
2N1969	2N2996.
2N1970	2N1100.
2N1971	2N2140.
2N1972	MPS6530, 2N2219.
2N1973	BC141/10, BFX85, BSX46-46/10, MPS6530, 2N2219.
2N1974/5	BC141/6, BF109, BFX85, BSX46-46/6, 2N3489-3498-4410.
2N1983	BCY65, BF177, BFX85, BSY92, MPS6530, 2N2218.
2N1984	BFX85, BFY33, BSX95, BSY10-91, MPS6530, 2N2218.
2N1985	BFX84, BFY33, BSX95, BSY91, MPS6530, 2N1893.
2N1986	BFY65, MPS6530, 2N2218.
2N1987	BFY65, BSX45, MPS6530, 2N2218.
2N1988/9	MPS6530, 2N1893-2243A.
2N1990	BFY45-65, BSW66-69, BSX21, 2N696/69/7-4410.
2N1990N/S/R	BFY45.
2N1991	BFX29, MPS6535, 2N1132-2905.
2N1992	2N2369A-3011-4264.
2N1993	2N1302-1306.
2N1994/5	2N1306-1308.
2N1997/8	2N1307.
2N1999	2N1997.
2N2000	ACY24, ASY48.
2N2002/3	2N2944-2945-4125.
2N2004/5	MPS404A, 2N2944-2945.
2N2006/7	MPS6516, 2N2944-2945.

2N2008	2N2990-5401.
2N2015/6	2N3715.
· 2N2017	MPSA05.
2N2018/9	2N3738-3996.
2N2020/1	2N3738.
2N2032	2N1209-4914.
2N2033	2N3420.
2N2034	2N3421-4238.
2N2035/6	2N4911.
2N2038	MPSA20.
2N2039	MPSA06.
2N2040	MPS6573.
2N2041	MPSA06.
2N2045	AF106, GMO760, 2N2635-2955.
2N2046	BF114, BFX85, BFY46, MPS6531.
2N2060/A/B	MPSA06.
2N2061	AF149, 2N2836.
2N2061A	AD130-138-149-150IV, TI3027, 2N2836.
2N2062A	AD131IV-149-149IV, TI3027, 2N2836.
2N2063	AD130/3-131/3/5-138, 2N2836.
2N2063A	AD131IV-138-149,-TI3027.
2N2064	AD130/4, OC26, 2N2836.
2N2064A	AD132III, ASZ15, TI3027, 2SB472.
2N2065 A	AD132III, ASZ15, TI3027-3030, 2SB472.
2N2066	AD131/4-132-138/50-149, GFT4012, SFT250, 2N1532.
2N2066A	AD132III-138/50, ASZ15, TI3022-3030, 2SB472, 2N1532.
2N2067	NKT404, 2N1536-2553.
2N2068	NKT403, 2N1531-2555.
2N2069	2N1539.
2N2070	2N1541.
2N2071	MP1539.
2N2072	MP1541.
2N2084	AFY11.
2N2085	2N1304.
2N2086/7	2N2243A-4410.
2N2089	AF130-178-194, 2N2188-2495.
2N2090	AF131-178-195, 2SA156, 2N2188-2190-2495.
2N2091	AF125-126-127-133-136-137-185-197, 2SA240,
2N2092	2N2188-2190.
	AF127-137-185, NKT613F, 2SA240, 2N2189-2192.
2N2093	AF117C-127-137-185, 2SA240, 2N2189-2193.
2N2095	2N2999.
2N2096/7	2N2997.
2N2098	2N2999.
2N2099	2N2997.
2N2100	2N2997.
2N2101	2N3487.
2N2102	BFR22, BSX46/6-17/0-04, 40412.
2N2102A	2N4410.
2N2104	2N2905.
2N2105	2N2904A-2905.
2N2106/7/8	MPS6530, 2N2987-4238.
2N2109 to 14	2N3846.

2N2116 to 19	2N346.
2N2123 to 26	2N4002.
2N2130 to 33	2N4002.
2N2137	AD130IV, AUY19IV, 2N1038-2552.
2N2137A	AD130IV-150III/IV, AUY19IV, 2N1038.
2N2138/A	AD149/IV, AUY19IV, 2N1038-2552.
2N2139	AD131IV, AUY19IV, 2N1038-2554.
2N2139A	AD131/V-138/50-149, AUY19IV, 2N1038-2554.
2N2140/A	AD132IV, AUY20IV, 2N1038-2555.
2N2141/A	AD163IV, AUY34IV, 2N1038-2555.
2N2142/A	AD130V, AUY19V.
2N2143/A	AD149V, AUY19V.
2N2144/A	AD131V, AUY19V.
2N2145/A	AD132V, AUY20V.
2N2146/A	AD163V, AUY34V.
2N2147	AD149-167, 2N1907-2832.
2N2148	AD149, 2N1908-2832.
2N2161	2N4400.
2N2162	2N2945-2946-4125.
2N2163/4	MPS3640, 2N2944-2945.
2N2165	2N2945-2946-4125.
2N2166/7	MPS3640, 2N2944-2945.
2N2168	2N2997.
2N2169/70	2N2996.
2N2171/2	2N1376-1377.
2N2175/6/7/8	MPS3639, 2N2944-2945.
2N2180	2N2635.
2N2181	MPS3638, 2N2944-2945.
2N2182	MPS3638, 3N108-111.
2N2183	MPSL08, 2N2944-2945.
2N2184	MPSL08, 3N108-111.
2N2185	2N2944-2945-4125.
2N2186/7	2N4125, 3N108-111.
2N2188/9	AF106, 2N3323.
2N2190/1	2N3323.
2N2192	BFX85, BSX45, MPS6531, 2N2219A-2243A.
2N2192A/B	BFX85, MPS6531, 2N2243A.
2N2193	BFY55, BSX45, MPS6531, 2SC708, 2N2218/A-2243A.
2N2193A	BSW53, BSX46, BSY46-85, MPS6531, 2N2243A.
2N2193B	MPS6531.
2N2194	BSX45, BSY71, MPS6530, 2N2218A.
2N2194A/B	MPS6530.
2N2195	BFY50-70, MPS6530.
2N2195A/B	BFY50, MPS6530.
2N2196	2N3766.
2N2197	BSX45, 2N3766.
2N2198	MPSA06.
2N2201/2/3/4	MPSL01, 2N4239.
2N2205	2N706-835-4264.
2N2206	MPS3640. 2N2410-3015.
2N2210	ASZ15, AUY21-28, 2N2075.
2N2214	MPS706, 2N706.
2N2216	2N3498.
2N2217	BCW77/16, BSX45, MPS6530, 2N2218-2219.

2N2218	BCW77/16, BSW51, BSX45-60, MPS6530, 2SC479, 2N2219.
2N2218A	BCW78/16, BSW53, MPS6530, 2N2219.
2N2219	BCW25-77/16, BSW52, BSX45-61-74, BSY34, MPS6530.
2N2219A	BCW78, BSW54, MPS6530.
2N2220	MPS6530, 2N2221-2222.
2N2221	BCW73/16, BSW61, BSX34-45, MPS6530, 2N2220-2222.
2N2221A	BC474/16, BSW63, MPS6530.
2N2222	BCW25-73/16, BFY34, BSW62, BSX45, BSY44, MPS6531, 2N2221.
2N2222A	BCW74, BSW64, BSX45A, MPS6531.
2N2223/A	MPSA06.
2N2224	MPS6530, 2N2218-2219.
2N2225	2N1143.
2N2230	BUY12-56/4.
2N2236	BSX73, BSY34, MPS834, 2N2193-2218-2243A-3053.
2N2237	BSX73, BSY58, MPS834, 2N2192-2218-2243A-2292-3053.
2N2239	MPSA10, 2N3766.
2N2240	2N699-2243A-4264.
2N2241	2N1890-2243A-4264.
2N2242	MPS3646, 2N2369/A.
2N2243	BSW65, MPS6530, 2N2219.
2N2243A	40412, 2N4410.
2N2244 to 55	MPS834, 2N929-930.
2N2256	BSX20, 2N4265.
2N2257	2N4265.
2N2258	2N964-972-4265.
2N2159	2N964-972.
2N2266/7/8/9	2N2145.
2N2270	MPS6530, 2N2102-2243A-3053.
2N2271	AC153, MFS6522, 2N404.
2N2272	MPS834, 2N914.
2N2273	AF106-121, AFY12.
2N2274	MPS3638, 2N2944-2945.
2N2275	MPS3638.
2N2276 to 79	MPS404.
2N2280/1	MPS3640.
2N2288	2N1046-1907.
2N2291	2N1907.
2N2294	2N1907.
2N2297	BFY50-55-56, BSY46-83, MPS6530, 2SC708, 2N2243A-3552.
2N2303	BFX87, 2N2801-5086.
2N2304	2N4910.
2N2305	2N4913.
2N2308	2N4912.
2N2309	MPS3708, 2N696-697.
2N2310	2N698-2243A.
2N2311	MPSA05, 2N698-2243A.
2N2312	MPSA06, 2N699-2243A.
2N2313	MPSA05, 2N1889-2243A.

2N2314	MPSA06, 2N698-2243A.
2N2315	MPSA20, 2N699-2222-2243A.
2N2316	MPSA20, 2N1893-2243A.
2N2317	2N1613-4410.
2N2318 to 20	MPS834, 2N834-3014.
2N2330/1	2N2432-4124.
2N2332/3	MPSL08, 2N2944-2945.
2N2334/5	MPS404, 2N2944-2945.
2N2336/7	2N2944-2945-5086.
2N2338	2N3713.
2N2339	2N1049-4910.
2N2349	MPS3706.
2N2350/A	MPS6531, 2N2192-2243A.
2N2351/A	MPSH04, 2N2193/A-2243A.
2N2352/A	MPS6530, 2N2194A-2243A.
2N2353/A	MPS6531, 2N2243/A.
2N2354	MPS2712, 2N1302.
2N2356/A	MPS4265, 3N76-79.
2N2360	AF139, 2N2997-3283.
2N2361	2N2997-3284.
2N2362	2N3284.
2N2363	2N2997.
2N2364/A	2N2243A-4410.
2N2368	BSX20-92, BSY21, MPS6530, 2SC689, 2N2369A-3227.
2N2369	BSX20-93, BSY19-21-63, MPS2369, 2SC689, 2N2369A-3224.
2N2369A	BSS17, BSY63, 2N3227.
2N2370 to 3	2N2944-2945.
2N2374/5	2N404.
2N2376	2N404-1193.
2N2377	MPS3394, 2N2944-2945.
2N2378	MPS6522, 2N2944-2945.
2N2380	2N1893-2243A-3903.
2N2380A	2N2217-2219-3903.
2N2383/4	2N4914.
2N2386	E270.
2N2387/8	BC237A, MPSA20.
2N2389	BSX45/6, MPSA05.
2N2390	MPSA05.
2N2391/2	MPS6522.
2N2393/4	MPS3703.
2N2395	MPSA20.
2N2396	BC140, BSY46, MPSA20, 2N2221A.
2N2397	MPS706, 2N2369A-3011.
2N2398/9	2N2997-3284.
2N2400	2N711-964.
2N2401	2N711A-964.
2N2402	2N711B-964.
2N2403/4	2N4400.
2N2405	2N1893-2243A-4410.
2N2410	BFY51, BSX45/10, 2SC479, 2N2218-3015-3903.
2N2411	BC177A-178VI, BCY78VII, 2N4124, BC177A-178V
2N2412	BC177A-178VI, BCY78, 2N4124.

2N2413	BC107, MPSH32.
2N2414	2N2060.
2N2417/A/B	2N3980.
2N2418/A/B	2N3980.
2N2419/A/B	2N3980.
2N2420/A/B	2N3980.
2N2421/A/B	2N3980.
2N2422/A/B	2N3980.
2N2423	2N456A-3146.
2N2424	MPS404A, 2N2944-2945.
2N2425	MPSA70, 2N2945-2946.
2N2427	MPS3694, 2N929-930.
2N2428	AC122-125-151-162-170, BC261, OC305, 2SB459, 2N1377-1381.
2N2429	AC125-151-163-171, BC151, OC305, 2N1377-1381.
2N2430	AC127-152-176-181-186, 2N1304-1605.
2N2431	AC117-128-153-153IV/V/VII, 2SB770A, 2N1377-1381.
2N2432	BCY59, 2N4123.
2N2432A	MPS3694.
2N2433/4	2N4400.
2N2435	2N4410.
2N2436	BC110, 2N4410.
2N2437/8/9	2N1893-2243A-4410.
2N2440	2N4410.
2N2443	40412, 2N1890-2243A-4410.
2N2444/5	2N456A-3146.
2N2447/8	AC122/30-125-151, 2N1187-1309.
2N2449	AC122-125-151, 2N1307.
2N2450	AC122-125-151, 2N652-1307.
2N2451	2N2635.
2N2453/A	MPS6530, 2N3680.
2N2456	2N999.
2N2460 to 66	MPSA06.
2N2472/3	2N2432-4410.
2N2474	2N2432-5227.
2N2475	BSX20, 2N3013-4265.
2N2476	BSW51, BSX45, 2N2218-2539-3015.
2N2477	BSW52, BSX45, 2N2218-2219A-2537-3015.
2N2478	BSW51, 2N2537-3015-4410.
2N2479	MPS6530, 2N3252.
2N2480	MPS6576, 2N2060.
2N2480A	MPS6576, 2N2639-2640.
2N2481	BSX20, 2N4265.
2N2482	2N797.
2N2483	BC107B, BCY58-65-65EVII, BFY27, MPSA05, 2SC648, 2N2484.
2N2484	BC107-107B, BCY59-65EIX, BFX93A, MPSA05, 2SC648, 2N2483.
2N2484A	MPSA05.
2N2487/8/9	2N2996.
2N2494	AF106-190, 2N2996.
2N2496	2N2996.
2N2497 to 500	E270.
2N2501	BSY19, 2N3014-5223.

2N2509	C407.
2N2514/5/6	MPSA05.
2N2517/8/9	2N4410.
2N2520/1/2	MPSA05, 2N929-930.
2N2523	MPS6530, 2N929-930.
2N2524	MPS6530, 2N930.
2N2526	AU105.
2N2527/8	AU103-105.
2N2529 to 34	MPS6576.
2N2535/6	2N1038-2565.
2N2537/8	BSX73, BSY34, 2N2218-3015-3903.
2N2539/40	BSX49, 2N2222/A-3015-3903.
2N2541	2N1038.
2N2551	2N5401.
2N2552 to 67	2N1038.
2N2571/2	2N3303.
2N2580	BUY28.
2N2581	BUY28, 2N3847.
2N2582/3	2N3847.
2N2586	BC107A, BCY66, BFY27, MPS6528, 2N2484.
2N2588	2N1038-2188.
2N2590 to 3	MPSH54, 2N2604-2605.
2N2595/6	MPSH54, 2N2604-2605.
2N2597	MPSH54, 2N736.
2N2598/9/0	2N2604-2605-5400.
2N2601/2/3	2N2604-2605-5086.
2N2604	BC177VI, BCY79IX, 2N2604-2605-4289-5086.
2N2605	BC177A, BCY79X, 2N4289-5086.
2N2605A	2N2605-5086.
2N2606 to 9	E176.
2N2610	MPSA20, 2N1149.
2N2613	AC126-151-151VI-160-160G, 2SC459, 2N404-2429.
2N2614	AC126-151-160-160G-163, 2N404-2429.
2N2616	MPS3563, 2N917-918.
2N2617	2N2944-2945-5400.
2N2618	MPS6532.
2N2632	2N3421-3487.
2N2633	2N3421-3488.
2N2634	2N3421-3489.
2N2639/40	MPSA20.
2N2641	MPSA20, 2N2639.
2N2642	MPSA20.
2N2643	MPSA20, 2N2642.
2N2644	MPSA20, 2N2643.
2N2645	MPSA05, 2N1711-2219.
2N2646	TIS43.
2N2648	2N1377-1379.
2N2651	2N2501-3554-5223.
2N2652/A	MPSA05, 2N2639.
2N2654	AF121-178-201-202, 2N654.
2N2656	MPSH07.
2N2657	2N2151-4238.
2N2658	2N2151-4239.
2N2659 to 70	2N1038.

2N2673 to 76	MPS6576.
2N2677	MPS6573.
2N2678	MPS6574.
2N2692/3	BCY59-70, 2N4123.
2N2694	BC108A, BCY59, 2N4123.
2N2695	BC328, BCY78, 2N2907-4126.
2N2696	BCY78, V763, 2N2837-2906-2907-4126.
2N2697/8	2N3998.
2N2699	2N964.
2N2706	AC126-128-131-151-152, 2N404.
2N2707	AC124-128-153, 2N2704.
2N2708	BFX62, BFY88-90, MPS6513, 2N918.
2N2709	2N2944-2945-4126.
2N2710	MPSH24.
2N2711	BC108-168A-238A, MPS2711, TIS98, 2SC458, 2N3709.
2N2712	BC168A-172-183-208-238/A/B, MPS2712-6565, 2SC458, TIS98, 2N2926-3710.
2N2713	BC168A-238A, MPS2713, TIS98, 2SC458, 2N2926-3709.
2N2714	BC168A-172-183-208-238/A, BSX38, MPS2714-6565, TIS98, 2SC458, 2N708-2926-3710.
2N2715	BC168A-238A, MPS2715, TIS98, 2SC458, 2N3663.
2N2716	BC168A-238A, MPS2716, 2SC458, 2N3663.
2N2717	2N2635.
2N2719	2N4265.
2N2720	MPSH04, 2N2639.
2N2721	MPSH04, 2N2639-2640.
2N2722	MPS3693, 2N2639.
2N2724	MPSA14.
2N2725	MPSA13.
2N2726/7	MPSU10.
2N2729	MPS3563.
2N2730	MP506.
2N2731	MP505.
2N2732	MP504.
2N2733	MP506.
2N2734	MP505.
2N2735	MP504.
2N2736	MP506.
2N2737	MP505.
2N2738	MP504.
2N2784	2N2369A-3010.
2N2786	AF139, AFY18, 2N2218.
2N2787/8	MPS6530, 2N2218.
2N2789	MPS6530, 2N2219A.
2N2790	MPS6530, 2N2218.
2N2791	MPS6530, 2N2221A.
2N2792	MPS6531, 2N2222A.
2N2795 to 99	2N2635.
2N2800	BC160/6, BFX87, 2N1132-2904-2905-3905.
2N2801	BC160/6-10, BFX87, 2N1132-3244-3905.
2N2802 to 07	MPS3702, 2N3350.
2N2808/9	MPS6543.

2N2810/A	MPS6568.
2N2811	2N3487-4301.
2N2812	2N3488-4301.
2N2813	2N3489-4301.
2N2814	2N3490-4301.
2N2815	2N4002.
2N2816 to 25	2N3846.
2N2831	MPSH24.
2N2832	2N1907-1908.
2N2835	AD139-148-152, 2N1038-2564.
2N2837/8	2N2904-2905-3905.
2N2840	2N489-491A.
2N2841 to 4	E270.
2N2845 to 8	MPS6531, 2N2218-2537-3015-4123.
2N2849	2N4410.
2N2850/1	2N3421-4410.
2N2852	2N3419-4410.
2N2853	2N3418-4400.
2N2854	2N4401.
2N2855	2N3420-4400.
2N2856	2N3418-4400.
2N2857	BFX62, BFY88-90, MPS6543, 2SC463, 2N918.
2N2858/9	2N4410.
2N2860	2N964.
2N2861/2	MPS6522.
2N2863/4	BFY51, MPS6565.
2N2865	BF180, BSY62-70, MPS3563, 2SC321.
2N2868	BSX22-45, MPSA06, 2N2243A-3252-4231.
2N2869	AD131/III-138/50-149, ASZ18, AUY22, MP2015, 2SB471, 2N301-3614.
2N2870	AD138/50, ASZ18, AUY21II-22-28, MP2016, 2SB340, 2N301A-3617.
2N2871	MPSH54.
2N2872	MPSL51.
2N2873	2N2997.
2N2875	2N3418.
2N2876	2N3632.
2N2877 to 80	2N3998.
2N2881	2N4235-5333.
2N2882	2N4236-5333.
2N2883/4	BFW17, MPSH20, 2N3553.
2N2885	2N847-4123.
2N2886	MPS3705, 2N696-697.
2N2887	2N4410.
2N2890	BDY12, BSW66, BSX47/16, 2N3421-4410.
2N2891	BD140, BDY12, BFY65, BSX47-47/10/16, 2N2991-3421-4410.
2N2892	BFX84, 2N3998.
2N2893	BFX85.
2N2894	BCY78, V763, 2N5226.
2N2894A	2N5226, SK3114.
2N2895	2N4410.
2N2896	2N5550.
2N2897	C762, 2N4123.
2N2898	2N4410.

2N2899	2N5550.
2N2900	C762, 2N4123.
2N2901	2N4265.
2N2903/A	MPS6566, 2N2639-2640.
2N2904	BCW80/10, 2N2905, SK3025.
2N2904A	BC161/10, BCW80/10, 2N2904-2905, SK3025.
2N2905	BCW80/16, SK3025.
2N2905A	BC161/10, BCW80/16, 2N2905, SK3025.
2N2906/A	BC212L, BCW76/10, 2N2907, SK3114.
2N2907	BC212L, BCW76/16, SK3114.
2N2907A	BC161/10, BCW76/16, 2N2907, SK3114.
2N2909	2N2193A-2343A-4123.
2N2910	MPS6530, 2N2639-3409.
2N2911	2N3421-3766.
2N2913/4	MPS6576, 2N2643.
2N2915/A	MPS6576, 2N2920.
2N2916/A	MPS6576, 2N2920.
2N2917/8	MPS6576, 2N2977.
2N2919/A	MPSA05, 2N2920.
2N2921	BC107B-168A-182/B-207-237, MPS6512-6565, TIS98, 2N2926/B, SK3020.
2N2922	BC108-168A-172-183-209-238, MPS6512-6523, TIS98, 2N2926/R, SK3020.
2N2923	BC108C-168-172C-173-183/C-238, MPS2923, TIS98, 2N2926/O-3710, SK3020.
2N2924	BC108C-168A-172/C-183/C-238, MPS2424-2924, TIS98, 2N3711, SK3020.
2N2925	BC108C-168B-172/C-183/C/LB-238, MPS2925, TIS98, SK3019.
2N2926	BC109B-172-183-184B-238, MPS2926, TI2926, TIS98, SK3018.
2N2926B/R	BC168A.
2N2926Y/O	BC168A.
2N2926G	BC168B.
2N2927	MPS3638 SK3114.
2N2929	2N1141.
2N2930	2N3427, SK3010.
2N2931/2/3	MPS2712, SK3020.
2N2934/5	MPS3709, SK3020.
2N2936	MPSA10.
2N2937	MPSA10, 2N2639.
2N2938	2N4265, SK3122.
2N2939	2N4409.
2N2940	BC160. BCY12. BF178, MPS2940, 2N1132-4410.
2N2941	BF178, MPL201.
2N2942/3	2N2635.
2N2944	BC178IV, MPS3640, 2N2945; SK3114.
2N2944A	MPS3640, 2N2945, SK3114.
2N2945/A	BC178VI, MPS3638, OK3114.
2N2946/A	2N4402, SK3114.
2N2951	2N2218A. SK3122.
2N2953	AC116-117R-126-151-153-184, 2N1194-2431 SK3004.
2N2954	2N918-4124.
2N2955	2N2635.

2N2956	ASY48, BSX20, BSY18-21, 2SC321, 2N2635-3635.
2N2957	2N2635.
2N2958	2N2217-2219, SK3122.
2N2959	2N697, SK3122.
2N2966	2N2997-3283.
2N2967	MPS706.
2N2968/-71	MPS3638, 2N2944-2945, SK3118.
2N2974/3	MPSA10, 2N2977.
2N2974/5/6	MPSA10, 2N2979.
2N2977	MPSA10.
2N2978	MPSA10, 2N2979.
2N2979	MPSA10.
2N2980	2N2060A.
2N2981/2	MPSA06, 2N2223.
2N2983	2N5550.
2N2984	2N5551.
2N2985	2N5550.
2N2986	2N5551.
2N2987	BC300, MPSA06
2N2988	2N5550.
2N2989	BC300. MPSA06.
2N2990	2N5550.
2N2991	MPSA06.
2N2992	2N5550.
2N2993	MPSA06.
2N2994	2N5550.
2N2999	2N3283.
2N3009	BSX20, BSY18-21, MPS2369, 2SC321, SK3122.
2N3010	BSX19-20, BSY18-21, MPS2369, 2N2369A, SK3039.
2N3011	BSS12, BSX20, BSY19-62B-63, MPS6516, 2N2369, SK3039.
2N3012	BC178VI, BSY62-70, MM2894, MPS3640, V763, 2N2894, SK3114.
2N3013	BSX19-20-48, BSY21-63, MPS2369, SK3122.
2N3014	BSX20-48, MPS2369, SK3122.
2N3015	BSY34, 2N2218-4400.
2N3019	MPSA06, 2N2243A.
2N3020	BC300, MPSA06, 2N1893-2243A.
2N3021 to 26	2N5384.
2N3033/4	BF177.
2N3035	BC107A, SK3039.
2N3036	BFY4580, 2N3019-4410.
2N3037	2N1893-4410.
2N3038	MPSHO5, 2N1893.
2N3039/40	MPS3703.
2N3043 to 7	MPS6575.
2N3048 to 50	MPS3702.
2N3051	MPS2369.
2N3052	MPSH04.
2N3053	BSX45, BSY44, 2N2243A, SK3024.
2N3054	BDY13, BUY38-46.
2N3055	AM291. BD130, BDX10, BDY39, 2N3713, SK3027.

2N3055U	BDX60.
2N3056V	BDX61.
2N3056/A	2N2243/A-4410.
2N3057/A	2N2243/A-4410.
2N3058	MPS3639, 2944-2945, SK3114.
2N3059	MPS3640, 2N2944-2945, SK3114.
2N3060	BSV16, MPSU55, 2N2944/5, SK3114.
2N3061	MPSU55, 2N2944/5.
2N3062 to 5	MPSL51, 2N2944/5, SK3114.
2N3066	E202.
2N3067/8	E201.
2N3069	E232.
2N3070	E230.
2N3071	E201.
2N3072	BFX29, 2N2904/5-5086, SK3114.
2N3073	BCY70, 2N2906/7-5086, SK3114.
2N3074	AF106-109, 2SA230, SK3006.
2N3075	AF121-200, 2SA230, SK3006.
2N3077	MPSA05, 2N2484, SK3020.
2N3078	MPSA05, SK3020.
2N3079	2N3846.
2N3080	2N3847.
2N3081	MPSH54, SK3114.
2N3082/3	2N4265, 3N76-79, SK3039.
2N3084 to 9	E202.
2N3107	BFX85, MPSA05, 2N2243/A.
2N3108	BFX84, BFY44-55, BSX45-46/16, BSY46, MPS46, MPSA06, 2N1613.
2N3109	BFY51, MPSA06, 2N697.
2N3110	BFY50, BSX45-45/6/10, MPSA06, 2N2218-2243A.
2N3112	E270.
2N3114	BD115, 2N5550, SK3104.
2N3115/6	MPS6530/1, 2N2221/2, SK3122.
2N3117	BSX20-63, BSY19-21-63, MPS6528, 2N2484, SK3020.
2N3118	MPSH04.
2N3119	2N4410.
2N3120	BFX87, 2N2800-3905, SK3025.
2N3121	BCY70, 2N2906/7-3905, SK3114.
2N3122	MPS3705, SK3122.
2N3123	2N4401.
2N3125/6	2N456A-3146, SK3014.
2N3128	MPS2712, SK3020.
2N3129	MPS6576, SK3122.
2N3130	MPSA05, SK3122.
2N3131	2N4123.
2N3132	TI3031, SK3009.
2N3133	BSV15-6, 2N2904/5-3905, SK3025.
2N3134	2N2905-3906, SK3025.
2N3135	MPS6534, 2N3905, SK3114.
2N3136	MPS6534, 2N3906, SK3114.
2N3137	BFX55, BSY58, MPSH20, 2N2218, 3014, SK3039.
2N3138	2N3766.
2N3146/7	2N456A-3616.

2N3152	2N4410.
2N3153	2N2482-4265.
2N3162	MPS3693, 2N3411.
2N3167	MJ2267.
2N3168	2N3789.
2N3169/70	2N3790.
2N3175	MJ2267.
2N3176	2N3789.
2N3177/8	2N3790.
2N3179	MJ2267.
2N3180	2N3789.
2N3181/2	2N3790.
2N3187	MJ2267.
2N3188	2N3789.
2N3189	2N3790.
2N3190	2N3790.
2N3192	2N3789.
2N3193/4	2N3790.
2N3200	2N3740.
2N3201	2N3741.
2N3202	2N5333, SK3035.
2N3203	2N3740-5333, SK3035.
2N3204	2N3741-5333.
2N3205	2N5333.
2N3206	2N3740-5333.
2N3207	2N3741-5333.
2N3208	2N3740-5333, SK3035.
2N3209	MM2894, MPS3640, 2N3576, SK3118.
2N3210/1	2N4123, SK3122.
2N3217/8	MPS3638, 2N2944/5, SK3114.
2N3219	2N2944/5-3905, SK3114.
2N3220/1	2N3767.
2N3222/3	2N3766.
2N3224	MPSL51, 2N2905-3498.
2N3225	MPSL51, 2N3498.
2N3226	2N4913.
2N3227	2N2369A-3011-5224, SK3039.
2N3232	2N3715.
2N3233	2N4347.
2N3235	2N3715.
2N3236	2N3716.
2N3238/9	2N3716.
2N3241	MPS3397, SK3020.
2N3241A	BC140-16, BSX75, MPS3397, 2N2222, SK3020.
2N3242	MPS3397, 2N730-2222, SK3020.
2N3242A	BC140-10, BSX75, MPSA20, 2N730-2222, SK3020.
2N3244/5	2N3905, SK3118.
2N3246/7	2N2484-3904. SK3020.
2N3248/9	MPS3640, 2N2894, SK3114.
2N3250	BCY70, MPS6516. SK3114.
2N3250A	BCY70, 2N3905, SK3114.
2N3251	BC177, BCY71-79VII, 2N3250-4125, SK3114.
2N3251A	2N3250-3906, SK3114.
2N3252	BFY51-55, BSX45, BSY34, 2N2218-3250-4400.

2N3253	BSX45-59, 2N2218A-4400.
2N3261	BSS10, BSX20, MPS2369, SK3122.
2N3263 to 6	2N4002.
2N3268	MPS6575, 2N337, SK3020.
2N3277/8	E176.
2N3279/80	2N2997.
2N3281 to 6	2N2996.
2N3287	BSX20, MPS6569, 2N2944/5, SK3018.
2N3288	BSX20, MPSH24, 2N2944/5, SK3018.
2N3289	BSX20, MPS6570, 2N918, SK3019.
2N3290	BSX20, MPS918, SK3019, [
2N3291	BF173, BSX20, MPS6540, 2N918, SK3019.
2N3292	BSX20, MPS6540, 2N918, SK3019.
.2N3293/4	BSX20, MPS6569, 2N918, SK3019.
2N3295	BSW51, MPSH04, 2N2217/8/9.
2N3296	MPSH04.
2N3298	2N5222.
2N3299	MPS6538, 2N2218-2537-3015.
2N3300	BSX61-73, BSY34, MPS6531, 2N2218.
2N3303	2N4265
2N3304	BSX20, MPS3639, 2N2894.
2N3305/6	MPSA70, 2N2907, SK3114.
2N3307	MPS6518, SK3118.
2N3308	2N5208, SK3118.
2N3309/A	BSW51, 2N2218-3553.
2N3310	MPS6546.
2N3317	2N2944/5-4125, SK3114.
2N3318/9	MPS3640, 2N2944/5, SK3114.
2N3320	2N964.
2N3321	2N964.
2N3326	MPS6530, 2N2218A.
2N3328 to 32	E176.
2N3337	MPS6569, 2N2883-3287.
2N3338	MPS6570, 2N2883-3289, SK3117.
2N3339	MPS6570, 2N2883-3288, SK3117.
2N3340	2N4124, SK3122.
2N3341	2N2604/5-4125, SK3114.
2N3342/3	MPS3638, 2N2944/5.
2N3344	2N2944/5-4125, SK3025.
2N3345/6	2N2944/5-4402, SK3025-3114.
2N3347 to 9	2N3350-4402.
2N3350	2N4402.
2N3351/2	2N3350-4402.
2N3365	E202.
2N3366/7	E201.
2N3368	E232.
2N3369	E230.
2N3370	E201.
2N3375	BLY59.
2N3376	E176.
2N3378	E270.
2N3380	E176.
2N3382	E176.

2N3384	E175.
2N3386	E174.
2N3388	2N4410, SK3045.
2N3389	2N2551, SK3045.
2N3390	BC108/C-130C-168-173C-183C/LC-238C, MPS6521-6565, TIS98, SK3020.
2N3391	BC108-173-183-238/B, MPS6515-6565, TIS98, SK3020.
2N3391A	BC108-109-148B-169B-173-184/LB-238B-239, MPS6520-6565, TIS98-SK3020.
2N3392	BC108-109-130A-168/A-172-183-238/A, MPS3392-6565, 2SC458, 2N2926/Y-3711, TIS98, SK3020.
2N3393	BC108-130A-167A-172-183-208-238, MPS3393-6565, TIS98, 2SC458, 2N2926/0-3710, SK3020.
2N3394	BC108-130A-168A/B-172-183-208-238, MPS3394-6565, TIS98, 2SC458, 2N2926/R-3709, SK3020.
2N3395	BC108-130-168A/B/C-172-183-208-238, MPS3395-6565, TIS98, 2SC458, 2N371/SK3020.
2N3396	BC108-130-168A/B/C-183, MPS3396-6565, TIS98, 2SC458, 2N3710, SK3018.
2N3397	BC108-130-168A/B/C-183, MPS3397-6565, TIS98, 2SC458, 2N3710, SK3020.
2N3398	BC108/A-130/A-168A/B/C-183, MPS3398-6565, TIS98, 2SC458, 2N3710, SK3020.
2N3399	AC139-189, AF139, GN0290, 2N2996.
2N3401	MPS3638, 2N2944/5, SK3114.
2N3402	BC108-183-338, MPS6513, TIS90, 2SC458, 2N3405-5449, SK3020.
2N3403	BC108-172-183-238-338, MPS6512-6515, TIS90, 2SC458, 2N3405-5449, SK3020.
2N3404/5	BC337, MPS6515, TIS90, 2N3405-5449, SK3020.
2N3407	MPS6541, SK3019.
2N3409 to 11	MPS6530, 2N2639.
2N3413	2N5401.
2N3414	BC108A-183-338, MPS6513-6565-2N3704-3710-5449, SK3020.
2N3415	BC108B-183-338, MPS6515-6565, 2N3704-3711-5449, SK3020.
2N3416	BC107A-141-182/LA-337, MPS6515-6566, 2N3704-5449, SK3020.
2N3417	BC107B-182/LB-337, MPS6515-6566, 2N3704-5449, SK3020.
2N3418	BFX84, MPSH04.
2N3419	2N4410.
2N3420	MPSH04.
2N3421	2N4410.
2N3423/4	MPS918, 2N2639.
2N3425	MPSH32, 2N3014.
2N3426	2N3303-4265.
2N3427/8	2N1377, SK3004.
2N3429	2N3713.
2N3430	2N3714.
2N3435	MPSH05.
2N3436	E203.

2N3437	E202.
2N3438	E201.
2N3439	2N5058.
2N3440	MPSU10, SK3021.
2N3441	BUY14, 2N3442.
2N3442	BDX11, BUY12, SK3079.
2N3444	BFY55, BSY45-61, BSY46, MPSH04.
2N3445	2N3715, SK3027.
2N3446/7	2N3715, SK3027.
2N3448	2N3716, SK3027.
2N3450	2N4410.
2N3451	MPSL07, SK3114.
2N3452	E202.
2N3453/4	E201.
2N3455	E202.
2N3456/7	E201
2N3458	E232.
2N3459	E230.
2N3460	E201.
2N3462	MPS6573, SK3020.
2N3463	MPS6576, SK3020.
2N3465/6	E202.
2N3467	BC160, 2N2905-3905.
2N3468	2N2905-3905.
2N3469	2N3420, SK3024.
2N3476	BUY12-56/4, 2N3442.
2N3478	BF173, MPS6548, TA2606, SK3019.
2N3485/A	BCW76/10, SK3114, 2N2906A-2907.
2N3486/A	BCW76/16, SK3114, 2N2907/A.
2N3493	2N4252-4265, SK3039.
2N3494	MPSU56, SK3025, 2N2605.
2N3495	2N2605-5400.
2N3496	MPSU56, 2N2605.
2N3497	2N2605-5400.
2N3498	2N697-698-4410.
2N3499	2N4410-5058.
2N3500/1	2N2243A-5550, SK3104.
2N3502	BC161, 2N2905-3602-3906, SK3025.
2N3503	BC161-161/16, MPSA55, 2N2905/A, SK3025.
2N3504/5	BC161-161/16, 2N2907-3906, SK3114.
2N3506	2N2989-3903, SK3024.
2N3507	MPSH05, 2N2989, SK3024.
2N3508	2N4123, SK3039.
2N3509	2N4123, SK3020.
2N3510	2N4123, SK3122.
2N3511	2N4264, SK3039.
2N3512	2N2218-2537-3015-4400.
2N3513	MPSA10, 2N2639-2640.
2N3514/5	MPSA10.
2N3516	MPSA05, 2N2639.
2N3517	MPSA05.

2N3518	MPSA05, 2N3046.
2N3519	MPS6512.
2N3520	MPS6512, 2N3043.
2N3521	MPSA05, 2N2643-3043.
2N3522	MPSA05, 2N2643.
2N3523	MPSA05.
2N3524	MPSA05, 2N2639-2640.
2N3526	2N4410.
2N3527	MPS3705, 2N2944-2945-SK3019.
2N3543	BDY61.
2N3544	MPS6511, 2N918, SK3039.
2N3545	2N4126, SK3114.
2N3546	MPS3639, 2N2894-3576, SK3114.
2N3547/8	MPSA55, SK3114.
2N3549	MPSA55, 2N2604-2605, SK3114.
2N3550	MPSA55, 2N2944-2945, SK3114.
2N3553	BFW47, BSX60.
2N3554	MPS6530.
2N3563	MPS3563, TIS62, SK3018.
2N3564	MPS6541, SK3018.
2N3565	BC108-130-172-183-208-238, MPS6514-6565, TIS98, 2SC485, SK3020.
2N3566	MPS6514, 2N5449, SK3020.
2N3567	MPS6530, SK3024.
2N3568	BC337, MPSA06, SK3020.
2N3569	MPSA05, MPS6531, SK3020.
2N3570	BFX89, SK3018.
2N3571	2N3570, SK3018.
2N3572	BFX89, BFY88, 2N3570, SK3018.
2N3573/4	E270.
2N3575	E176.
2N3576	2N4126.
2N3577	2N4410.
2N3578	E270.
2N3579	MPSH04, 2N2604-2605, SK3114.
2N3580	MPSH04, 2N2605, SK3114.
2N3581	MPSH55, 2N2605, SK3114.
2N3582	MPSA70, 2N2605, SK3114.
2N3585	BDY94, 2N3767.
2N3586/7	MPS6530.
2N3588	AF178-201, 2SA239, 2N2495, SK3006.
2N3592/3/4	2N5551.
2N3597	2N3767-4002.
2N3598	2N3446-4002.
2N3599	2N3447-4002.
2N3600	BFX62, BFY88, MPS3563, 2N918, SK3019.
2N3605	BSX80, BSY62, MPS3646, 2N914, TIS48, SK3019.
2N3605A	BSX20, BSY63, MPS3946, 2SC321.
2N3606	BSY62, MPS3646, TIS48, SK3019.
2N3606A	BSX20, BSY63, MPS3646-3946, 2SC321.
2N3607	BSX20-80, BSY62-70, MPS3946, TIS48, 2SC321, SK3019.

2N3611	AUY21IV, TI3027, SK3009.
2N3612	AUY21IV, SK3009.
2N3613/4	AUY21V, TI3027, SK3009.
2N3615	ASZ16, AUY22IV, TI3027-3031, SK3009.
2N3616	ASZ15, AUY22IV, 2N456A-3146, SK3009.
2N3617	ASZ16, AUY22V, TI3027-3030, SK3009.
2N3618	AUY22V, 2N456A-3146, SK3009.
2N3632	BLY23-60.
2N3633	MPS3640, SK3039.
2N3634/5	2N5400.
2N3636/7	2N5401.
2N3638/A	MPS3638, TIS50, 2N5447, SK3025.
2N3639	MPS3639, TIS53, SK3114.
2N3640	MPS3640, TIS54, SK3114.
2N3641	BCW36, MPS6530, 2N5449, SK3018.
2N3642	BC337, BCW36, MPS6530, 2N2219-5449, SK3122.
2N3643	BFR41, MPS6531, 2N2219-5449, SK3024.
2N3644	BC327, MPS6516, TIS50, 2N5449, SK3114.
2N3645	MPSA55, TIS50.
2N3646	BSX80, BSY34, MPS3646, TIS55, 2N4422, SK3019.
2N3647	2N3303-4123.
2N3648	2N3303-4123, SK3039.
2N3659	2N5058.
2N3660	2N3719.
2N3661	2N3720.
2N3662	BF200-224J, MPS3563, SK3039.
2N3663	BF224J, MPS3563, SK3018.
2N3665/6	2N2432.
2N3667	2N3715, SK3036.
2N3671	MPS6533, 2N2905, SK3025.
2N3672/3	MPS6533, 2N2905, SK3114.
2N3675	2N4238.
2N3676	2N4239.
2N3677	2N2944-2945-4125, SK3025.
2N3678	MPS3693.
2N3680	MPS6530.
2N3681	MPS3563, 2N3570.
2N3682	MPS6532, 2N918, SK3039.
2N3683	MPS6543, 2N3570.
2N3684/A	E231.
2N3685/A	E230.
2N3686/A-7/A	E201.
2N3688/9	MPSH24, SK3039.
2N3690	MPSH20, SK3018.
2N3691	MPS6512, TIS98, SK3018.
2N3692	MPS6513, SK3020.
2N3693	MPS3693, SK3039.
2N0094	MPS3694, SK3018.
2N3695	E270.
2N3696	E176.
2N3697/8	E270.
2N3700	BSX47/10, 2N4410, SK3045.
2N3701	BSX47/6, 2N4410.

2N3702	BC157-177-257A-261-308VI, MPS3702, 2SA565, SK3025.
2N3703	BC147-157-177-257VI-261-307VI-370B, MPS3703, 2SA565, SK3114.
2N3704	BC140-337/16, MPS3704, 2N2222, SK3024.
2N3705	BC140-337-337/16, MPS3705, 2N2221, SK3020.
2N3706	BC140-168-337/16-338, BSX75, MPS3706, 2N2222, SK3020.
2N3707	BC149-149B-167B-169B-173-173B-184-209-209B-237A-370, MPS3707-6520, 2SC458, SK3020.
2N3708	BC147A-167A-B-C-171A-207A-237A, MPS3708-6566, 2SC458, SK3020.
2N3709	BC147A-167-A-170-171A-207A-237A, MPS3709-6566, 2SC458, SK3020.
2N3710	BC147A-167A/B-171/A-182-207/A-237/A, MPS3710-6565, 2SC458, SK3020.
2N3711	BC107-147B-167A/B/C-171B-207B-237B, MPS3711-6566, 2SC458, SK3020.
2N3712	BD115, 2N5551, SK3045.
2N3713	BDY34/4, SK3036.
2N3714	BDY39/4, SK3036.
2N3715/6	BDY39/6, SK3036.
2N3717	BSY34.
2N3719/20	BSX62/6-10, 2N5333, SK3025.
2N3721	BC168A/B/C, MPS3721-3731, 2N3710, SK3020.
2N3722	BSX46/6-10, 2N3015-4400.
2N3723	BSX47/6-10, 2N3015-4410, SK3104.
2N3724	BSX60-73, BSY34-58, 2N3725-4123.
2N3725	BSX59-73, BSY34, MPS3725.
2N3726/7	MPS6518.
2N3728/9	MPS6530.
2N3731	AU106, SK3035.
2N3732	AUY22-28, SK3034.
2N3733	2N3632.
2N3734/A	MPS6532.
2N3735/A	2N4409.
2N3736/A	MPS6532.
2N3737	2N3725-4409.
2N3737A	2N4409.
2N3738	BSX59, BU126, SK3021.
2N3742	MPSU10, 2N5058.
2N3743	MPSU10.
2N3744 to 52	2N3996.
2N3762	2N3244-3905, SK3025.
2N3763	2N3905, SK3025.
2N3764/5	2N2907-3486-3905, SK3025.
2N3766	BD124, BDY13-13/10.
2N3767	BDY13/10.
2N3771	BDX41, 2N5301, SK3036.
2N3772	BDX40, 2N5302, SK3036.
2N3773	BD124, BDX50, BDY12, BUY12-56/4.
2N3774	2N4234, SK3025.

2N3775	2N4235, SK3025.
2N3776	2N4236, SK3025.
2N3777	2N4033-4236.
2N3778	2N4234, SK3025.
2N3779	2N4235.
2N3780/1	2N4236, SK3025.
2N3782	2N4234, SK3025.
2N3783/4	TIXM101, 2N5043.
2N3790	2N3789.
2N3791/2	2N3789.
2N3793	BC167-168/A-171-182-207-237, MPS6530, 2N3709, SK3020.
2N3794	BC167-168/A-172-183/L-208-238, MPS6531, SK3020.
2N3798	2N2605-5086, SK3114.
2N3799	2N2605-5087, SK3114.
2N3800/1	MPSA55, 2N3350-3352.
2N3802	MPSA55, 2N3347-3350.
2N3803	MPSA55, 2N3350/1.
2N3804 to 11	MPSA55, 2N3350.
2N3812 to 16	MPSA55, SK3114.
2N3817/A	MPSA55.
2N3819	E304, SK3118.
2N3820	E176, SK3116.
2N3821	E202.
2N3822	E210.
2N3823	E304.
2N3824	E114.
2N3825	BC109-173-184-209, MPS3398, SK3039.
2N3826	MPS3826, 2N4994, SK3018.
2N3827	MPS3827, 2N4995, SK3018.
2N3828	BC107-109-171-173-182-207, MPS6565, SK3122.
2N3829	2N3964-4125, SK3114.
2N3830	BFX34, 2N4409.
2N3831	BD131, 2N4409.
2N3832	BSY19, 2N4265, SK3039.
2N3836	2N4409.
2N3837	2N4410.
2N3838	BC107-172-182-207, MPS6565.
2N3839	BFX62, MPS6541, 2N3570/1.
2N3840	2N2945/6-5086, SK3114.
2N3841	MPSL51, 2N2945/6.
2N3842	2N2945/6-5400.
2N3843	MPS6512, SK3122, 2N3709.
2N3844	MPS6512, SK3122.
2N3844/A	2N3709, SK3122.
2N3845	BC107-109-171-173-182-207, MPS6512-6565, 2N3710, SK0009.
2N3845A	MPS6513, SK3039.
2N3850 to 3	2N3998.
2N3854	MPS6512, SK3039.
2N3854A	MPS6512/3, 2N3709, SK3039.
2N3855	BC107A-182, MPS6512-6566, SK3039.

2N3855A	BC238A, MPS6512, SK3039.
2N3856	BC107A/S-108-168B-182-238A/B, MPS6513-6566.
2N3856A	MPS6513, SK3039.
2N3857	MPSA55, 2N2944/5, SK3114.
2N3858	MPS6512, TIS98, 2N3710, SK3122.
2N3858A	BC182LA, MPS6566, TIS98, SK3122.
2N3859	MPS6513, TIS98, 2N3711, SK3020.
2N3859A	BC182LB, MPS6566, SK3122.
2N3860	MPS6514, TIS98, 2N3711, SK3039.
2N3862	2N4123, SK3122.
2N3863	2N3715, SK3036.
2N3864	2N3716, SK3036.
2N3866	BFX55, SK3024.
2N3867	2N3905-5333, SK3025.
2N3868	2N5086-5333, SK3025.
2N3877	BF178, BSS34, TIS98, 2N4410, SK3024.
2N3877A	BSS35, 2N4410, TIS98.
2N3878	BD124, BDY13/C, SK3021.
2N3880	MPS6543, SK3122.
2N3881	MPS6530.
2N3900	BC183LB, TIS98, 2N5088, SK3020.
2N3900A	BC184LB, TIS98, 2N5088, SK3020.
2N3901	2N5088, SK3020.
2N3903	BC167A-237A, BF195, BFY19.
2N3904	BC167A-237A, BF194/5, BFY19, BSY34, SK3024,
2N3905	BC212-251-307/A, BCY70, SK3114.
2N3906	BC212-251-307/A, BCY70, SK3025.
2N3907	MPS6576, 2N2915.
2N3908	MPSAQ6, 2N2916.
2N3909/A	E176.
2N3910	2N2944-5086.
2N3911/2	2N2944/5-5086, SK3114.
2N3913	2N2944/5-5086.
2N3914	2N2906-2944/5-5086, SK3114.
2N3915	2N2944/5-5086.
2N3921	E400.
2N3922	BFX62, BFY88, E401, 2N3933.
2N3923	2N5550, SK3045.
2N3924	BFW46, BFY99, 2N3553.
2N3925	2N3375.
2N3926	BLY57.
2N3927	BLY58.
2N3930	2N2605-3497-5401.
2N3931	2N5401.
2N3932	BFX62, BFY88, MPS6541.
2N3933	MPSH20, SK3039.
2N3934	E400.
2N3935	E401.
2N3941	2N2920.
2N3942/3/4	MPS6530, 2N2920.
2N3946	MPS6530, 2N2217-2219.
2N3947	2N2217-2219-3904.

2N3948	2N3904, SK3024.
2N3953	MPS6543.
2N3954/A	E400.
2N3955/A	E401.
2N3956/7/8	E402.
2N3959	MPS3640.
2N3960	BFX62. MPS3640.
2N3962	2N2605-3963-5086, SK3114.
2N3963	2N2605-5086, SK3114.
2N3964	BCY67, 2N2605-5087, SK3114.
2N3965	2N2605-5086, SK3114.
2N3966	E304.
2N3967/8	E202.
2N3969/A	E230.
2N3970	E111.
2N3971	E112.
2N3972	E113.
2N3973	2N4400-5449, SK3122.
2N3974	2N4401-5449, SK3122.
2N3975	2N3709-4400-5449, SK3122.
2N3976	2N3710-4401-5449, SK3024.
2N3977	MPS3640, 2N2944/5, SK3114.
2N3978	MPS3638, 2N2944/5, SK3114.
2N3979	2N2944/5-4402, SK3114.
2N3981	2N4400.
2N3982	2N4123.
2N3983	MPS6548, TIS62, SK3039.
2N3984/5	MPS6548, TIS63, SK3018.
2N3993/A	E174.
2N3994/A	E176.
2N3995	2N1195-2929.
2N4000	2N3019-4410.
2N4001	BSW66, 2N4410.
2N4006	MPS6522.
2N4007	MPS3638.
2N4008	2N2944/5-4125.
2N4013	BSX48, 2N3903, SK3122.
2N4014	BSX49, 2N2219-4409, SK3122.
2N4015	MPS6531, 2N3350.
2N4016	2N3350-5086.
2N4017	MPSH54, 2N3350.
2N4018	2N3350-5086.
2N4019	MPSA70.
2N4020	MPSA70, 2N3350.
2N4021/2	MPSA55, 2N3350.
2N4023	MPSA70, 2N3350.
2N4024/5	MPSA55, 2N3350.
2N4026	BC160, BCW76-10, MPSA55, 2N1132-2906A/7, SK3025.
2N4027	MPSA56, 2N2906A/7, SK3025.
2N4028	MPSA55, 2N2907, SK3025.
2N4029	MPSA56, 2N2907, SK3025.
2N4030	BC161-161/6, BCW80/10, BFX29, MPSA55, 2N2905, SK3025.

2N4031	BC161-10, BCW80-10, MPSA56, 2N2905, SK3025.
2N4032	BC161-16, BCW80-10, MPSA55, 2N2905, SK3025.
2N4033	BC161-16, BCW80-16, MPSA56, 2N2905, SK3025.
2N4034/5	2N3250-3905.
2N4036	BFR23, 2N2904/5.
2N4037	BFR24, 2N2904/5-4402.
2N4042	MPS6566, 2N3680.
2N4043	MPS6576, 2N3680.
2N4044	MPS6566, 2N3680.
2N4045	MPS6576, 2N3680.
2N4046	2N2218-3252-4400.
2N4047	2N4409.
2N4054/7	SK3103.
2N4058	BC178/9-206-253-258VI/A/B-263-309, BCY72, MPS6522, SK3118.
2N4059	BC177-204-250-258/A/B/VI-261-307-308B, MPS6516, SK3118.
2N4060	BC258A/VI-308VI, MPS6516, SK3118.
2N4061	BC258A/B/VI-308A, MPS6517, SK3118.
2N4062	BC258A/B-308B, MPS6518, SK3118.
2N4068/9	2N5551, SK3045.
2N4070	2N3448.
2N4072/3	2N2863, SK3018.
2N4074	MPS3564, SK3122.
2N4075	2N3764-3996.
2N4076	BD124, BDY13-13/10, 2SC830, 2N3996.
2N4078	2SB267, SK3052.
2N4079	AD161/2.
2N4080	2N5226, SK3118.
2N4081	MPS3563, 2N4252.
2N4082	E400.
2N4083	E401.
2N4084	E400.
2N4085	E401.
2N4086	MPS6514, TIS98, SK3020.
2N4087/A	MPS6515, TIS97, SK3020.
2N4088 to 90	E176.
2N4091/A	E111.
2N4092/A/3/A	E112.
2N4094/5	E111.
2N4099	2N3680-5210.
2N4100	2N5210.
2N4101	SK3502.
2N4104	2N2484, SK3020.
2N4105	SK3010.
2N4106	AC128-131-152-153, MPS2060, 2SB370A, SK3004.
2N4115	2N3996.
2N4116	BD124, BDY13-13/16, 2SC830, 2N3996.
2N4117/8/9	E201.
2N4121	2N3905-4423, SK3118.
2N4122	2N3906-4423, SK3118.
2N4123	TIS99, 2N3903, SK3020.

2N4124	BC171-182-207-237, TIS98, 2N3904, SK3020.
2N4125	2N3905-5447, SK3114.
2N4126	BC213-252-308, 2N3906-4061, SK3114.
2N4130	SK3036.
2N4134/5	MPS6540.
2N4137	2N4265.
2N4138	2N2432-4124.
2N4139	E305.
2N4140	2N4400, SK3122.
2N4141	2N4401, SK3122.
2N4142	2N4402, SK3114.
2N4143	2N4403, SK3114.
2N4150	2N3421.
2N4207	MPS3639.
2N4208/9	MPS3640, SK3118.
2N4210/1	2N4002.
2N4220/1	E202, SK3112.
2N4222/A	E203, SK3112.
2N4223	E210.
2N4224	E300.
2N4227	2N4400.
2N4228	2N4402.
2N4234	2N3905, SK3025.
2N4235	2N5086-5333, SK3025.
2N4236	MPSH54, 2N5333.
2N4237	SK3024.
2N4241	AD149, TI3027, 2SB471, SK3013.
2N4242 to 6	TI3027/8.
2N4247	TI3027/8, TIS47.
2N4248	BC213, 2N4058-5086, SK3114.
2N4249	BC212, 2N4058-5086, SK3118.
2N4250	BC214, 2N4059-5087-5987, SK3118.
2N4251	2N4265, SK3039.
2N4252/3	MPS6548.
2N4254/5	BF195, MPS6547, 2N4996/7, SK3039-3118.
2N4256	2N3904, SK3122.
2N4257	MPS3639, TIS53, SK3118.
2N4258	MPS3640, TIS54, SK3114
2N4259	BFX62, BFY88, MPSH32, 2N4252.
2N4260	MPS3638.
2N4264/5	BC238A, TIS55.
2N4269	2N2243A-5551, SK3045.
2N4270	2N5551, SK3045.
2N4274	2N4264-4419, SK3039.
2N4275	TIS47, 2N4265-4418, SK3039.
2N4284	MPS6516, 2N2944/5-4060.
2N4285	MPS6516, 2N4060.
2N4286	BC107-168/A-1/1-182-184LB-207-237-238C, MPS6515/6-6566, TIS98, 2SC458, SK3020.
2N4287	BC171-182-184LB-207-237-277, MPS6566, TIS98, 2SC458, SK3020.
2N4288	BC178/B-205-212LB-213-262, MPS6518, 2SC565, SK3114.

2N4289	BC177/B-307B, 2SA565, 2N4058-5086, SK3114.
2N4290	BC179-206-214-263, MPS6533, 2N3702-5447, SK3114.
2N4291	BC179-206-214/LA-263, MPS6534, 2N5447.
2N4292	BC173-184-209-224L-239, BFX59-62, MPS918-6518, TIS62, 2N1132.
2N4293	BC139-173-179-184-209-239, MPS918-6518, TIS62.
2N4294/5	2N4264.
2N4296/7/8	SK3021.
2N4302	E202.
2N4303/4	E203.
2N4313	MPS3640, 2N4423, SK3118.
2N4314	MPSH55.
2N4338/9	E201.
2N4340	E202.
2N4341	E231.
2N4342/3	E176.
2N4347	BDX12, SK3079.
2N4348	BDX51, SK3079.
2N4354	MPS4354, SK3025.
2N4355	MPS4355, SK3114.
2N4356	MPS4356, SK3114.
2N4357/8	2N2605-3494-5401.
2N4359	2N5086.
2N4360	E176.
2N4381	E176.
2N4382	E176, MPS3705.
2N4383 to 6	MPS3705.
2N4389	MPS3640, SK3118.
2N4390	BFR25, MPSL01, 2N5058.
2N4391	E111.
2N4392	E112.
2N4393	BSV68, E113.
2N4395/6	2N3715, SK3027.
2N4397	MPS6532, 2N4252.
2N4400	BFR39, 2N5449, SK3122.
2N4401	BFR41, 2N5449, SK3122.
2N4402	BC307VI, BFR79, 2N2904/5, SK3114.
2N4403	BC307A, BFR81, 2N2905, SK3025.
2N4409	BFR34.
2N4410	BSS35, SK3045.
2N4411	MPSLO8, SK3114.
2N4412	MPS3705, SK3025.
2N4412A	MPSH54, SK3025.
2N4413	MPS6533, SK3114.
2N4413A	MPSH55, SK3114.
2N4414	MPS6533, SK3025.
2N4414A	MPSH55, SK3025.
2N4415	MPS6533, SK3114.
2N4415A	MPSH55, SK3114.
2N4416/7	E304, SK3112.
2N4418/9	2N4264, SK3039.
2N4420/1/2	MPS3646, SK3118.
2N4423	MPS3640, 2N4421.
2N4424	BC182LB-337, MPS3711, SK3020.

2N4425	BC337, MPS3711, TIS90, SK3020.
2N4428	BFX55.
2N4432/A	MPS6532, SK3122.
2N4433/4	BF115, SK3018/3117.
2N4436	MPS6530, 2N5449, SK3122.
2N4437	MPS6531, 2N5449, SK3122.
2N4438/9	MPSU10.
2N4440	2N3632.
2N4445	E106.
2N4446	E109.
2N4447	E106.
2N4448	E109.
2N4449	2N4265, SK3039.
2N4451	MPS3640.
2N4452	MPS3638, SK3114.
2N4453	MPS3640, SK3118.
2N4856/A	E111.
2N4857/A	E112.
2N4858/A	E113.
2N4859/A	E111.
2N4860/A	E112.
2N4861/A	E113.
2N4867/8/A	E230.
2N4869/A	E231.
2N4872	MPS3640, SK3118.
2N4873	2N4265, SK3039.
2N4874/5	MPS6543.
2N4876	MPSH34.
2N4878	MPSH05.
2N4879	2N5210.
2N4880	2N5401.
2N4888/9	2N5401.
2N4890	BSV16-16/6, 2N3905.
2N4916/7	2N5208, SK3025/3114.
2N4918	BD132, TIP30.
2N4919	BD138, TIP30A.
2N4920	BD140, TIP30A.
2N4921	BD131, TIP29.
2N4922	BD137, TIP29A.
2N4923	BD139, TIP29A.
2N4924	2N4410, SK3045.
2N4925	2N5550, SK3045.
2N4926	2N5551, SK3045.
2N4927	MPSU10, SK3045.
2N4934	MPS6543, 2N4252, SK3039.
2N4935	MPS3826, 2N4252, SK3039.
2N4936	MPS3826, 2N5449, SK3039.
2N4937 to 42	MPS6534.
2N4943	2N4410, SK3024.
2N4944	2N4252-4400, SK3024.
2N4945	MPSA06, SK3024.
2N4946	2N4401-5449, SK3024.

2N4951	BC183L-337, MPS6530, 2N3705-5450, SK3020.
2N4952	BC184L-337, MPS6531, 2N3704-5449, SK3020.
2N4953	BC184LC-337, MPS6531, SK3020.
2N4954	BC183L-338, MPS6530, 2N5449, SK3020.
2N4960	MPSA05, SK3024.
2N4961	MPSA06, SK3024.
2N4962	MPSA05, SK3024.
2N4963	MPSA06, SK3024.
2N4964	MPS6516, 2N4060, SK3024.
2N4965	MPS6552, 2N4058, SK3114.
2N4966/7	MPSA20, TIS99, SK3020.
2N4968	MPS3397, TIS99, SK3020.
2N4970	2N4400, SK3122.
2N4971/2	2N3905, SK3114.
2N4973	2N5221.
2N4976	TIS39, 2N4875.
2N4977/8	E110.
2N4979	E112.
2N4980	2N4125.
2N4981	2N4402.
2N4982	MPSH54, SK3114.
2N4994	MPS3693, SK3122.
2N4995	MPS3694, SK3122.
2N4996	MPSH10, SK3018.
2N4997	MPSH11, SK3018.
2N5006	BD183.
2N5018	E174.
2N5019	E175.
2N5020	E176.
2N5021	E176.
2N5022	2N2386, SK3025.
2N5023	2N2386-4125, SK3025.
2N5024	2N4265, SK3039.
2N5027/8	MPS3646.
2N5029	2N4265.
2N5030	2N4264.
2N5033	E176.
2N5034/5	TIP33.
2N5036	TIP33A, 2N3055.
2N5037	BD181, TIP33A.
2N5040	MPS3702.
2N5041/2	MPSA70.
2N5045/6/7	E231.
2N5053	2N918, SK3018.
2N5055/6/7	MPS3640, 2N3829, SK3118.
2N5058/9	MPSU10, SK3045.
2N5066	2N2432-4124.
2N5073	2N5551.
2N5078	E300.
2N5083	2N3055.
2N5086	BC177-212/KAS-257A-261-307A, MPS6516, 2SA565, 2N4058, SK3114.

2N5087	BC159A-177-179-212/KBS-259A/B-261; MPS6516.
2N5088	BC169C-184KBS-237B, 2N5089, SK3020.
2N5089	BC169C-184KCS-239C, SK3020.
2N5103/4	E305.
2N5105	E300.
2N5106/7	MPS6518, 2N5399.
2N5114	E174.
2N5115	E175.
2N5116	E176.
2N5126	MPSH10, TIS98, SK3020.
2N5127	MPS918, TIS98, SK3020.
2N5128/9	MPS6514, 2N5451, SK3020.
2N5130	MPS3563, 2N5450, SK3018.
2N5131	MPS3646, 2N5451, SK3020.
2N5132	MP6539, 2N5451, SK3020.
2N5133	MPS2714, 2N5449, SK3018.
2N5134	2N4422-5224, SK3020.
2N5135	MPS2711, SK3020.
2N5136	MPS3706, SK3020.
2N5137	MPS6560, SK3020.
2N5138	MPS6516, 2N4061, SK3114.
2N5139	BC308VI, MPS6516, 2N3250, SK3114.
2N5140	MPS6513, 2N3250, SK3118.
2N5141	MPSL07, 2N3829, SK3118.
2N5142/3	MPS3638, 2N3829, SK3114.
2N5144/5	2N4400.
2N5146	MPS451, 2N4031.
2N5148	BSW10-65, BSX63, 2N4410.
2N5149	MPSL51, 2N4033.
2N5150	2N3019-4410.
2N5151	MPSL51, 2N4031.
2N5152	BSW10-65, BSX63, 2N4410.
2N5153	MPSL51, 2N4033.
2N5154	2N3019-4410.
2N5158	E107.
2N5159	E106.
2N5163	E300, SK3116.
2N5172	BC237A, MPS5172, 2N3711, SK3020.
2N5179	2N5224, SK3039.
2N5180	MPS6548, SK3018.
2N5181	MPSU04.
2N5182	MPS6548.
2N5183	2N5222, SK3020.
2N5184	BF178-257, MPSL01, SK3040.
2N5185	MPSL01, SK3040.
2N5186	2N4265, SK3122.
2N5187	2N424, SK3122.
2N5188	MPS6565, SK3024.
2N5189	BSS13, BSX45, MPS6565, 2N3053.
2N5190	TIP31, SK3054.
2N5191/2	TIP31A, SX3054.
2N5193	TIP32. SK3083.

2N5194/5	TIP32A, SK3083.
2N5196/7	E400.
2N5198	E401.
2N5199	E402.
2N5208	TIS37, SK3118.
2N5209	BC182KAS-237/A/B/C, SK3122.
2N5210	BC182KBS-237-237B/C, SK3122.
2N5219	BC238-238B/C-239B, 2N3708KS, SK3122.
2N5220	BC338-338/16/25/40, 2N3706KS, SK3122.
2N5221	BC338-338/16/25/40, 2N3703KS, SK3114.
2N5222	BF224J-254, SK3039.
2N5223	BC239B, 2N3708KS, SK3122.
2N5224	TIS48, SK3122.
2N5225	BC338-338/16/25/40, 2N3706KS, SK3122
2N5226	BC338-338/16/25/40, 2N3702KS, SK3114.
2N5227	2N4061KS, SK3114.
2N5228	TIS53.
2N5232	BC182LB, MPSA05, SK3122.
2N5232A	BC167B-182LB, SK3122.
2N5233/4/5	MPSA05, SK3024.
2N5236	MPS6511.
2N5240	BDY97.
2N5243	MPS3638, 2N4125.
2N5245	E304, SK3116.
2N5246	E305, SK3116.
2N5247	E304, SK3116.
2N5248	E300, SK3116.
2N5249	BC167B/C, MPS6514, SK3024.
2N5249A	MPS6528, SK3024.
2N5255/6	MPSA70.
2N5262	BFX34, BSS14, 2N4409.
2N5265/6	E176.
2N5267	E231.
2N5268/9/70	E232.
2N5272	2N4401.
2N5276	2N4264.
2N5277	E232.
2N5278	E305.
2N5281	2N5401.
2N5284	BDY90.
2N5288	BDY90.
2N5292	MPS3640.
2N5293/4	TIP31A.
2N5295	TIP31.
2N5296	TIP31, 2N3054S, SK3054.
2N5297/8	TIP31A.
2N5309/10	BC167A, 2SC458, 2N4401, SK3024.
2N5311	BC167B, MPS6514, 2SC458, SK3024.
2N5312/3/4	2N5386.
2N5316/7/8	2N5384.
2N5320	BSS15, BSV94.
2N5321	BSS16, BSV93.

2N5322	BSS17, 2N4036.
2N5323	BSS18, 2N4036.
2N5332	MPS3638.
2N5333	MPSL51.
2N5354	BC328, MPS3638, 2N3702, SK3114.
2N5355	BC213L-328, MPS368A, SK3114.
2N5356	BC214LC-328, MPS6534, SK3114.
2N5358/9	E202.
2N5360	E305.
2N5361	E304.
2N5362/3/4	E305.
2N5365	BC212L-327, MPS3638, SK3114.
2N5366	BC212LA-327-328-337, MPS6534, SK3114.
2N5367	BC212LB-327, MPS6534, SK3114.
2N5368	BC183, MPS6530, SK3122.
2N5369	BC184, MPS6531, SK3122.
2N5370	BC184C, MPS6530, SK3122.
2N5371	BC183, MPS6530, SK3122.
2N5372	BC213, MPS6533, SK3114.
2N5373	BC214, MPS6534, SK3114.
2N5374	BC214C, SK3114.
2N5375	BC213, MPS6533, SK3114.
2N5376	BC184, MPS6521.
2N5377	BC183, MPS6520.
2N5378	BC214, MPS6523, SK3114.
2N5379	BC213, MPS6522, SK3114.
2N5380	A5T3903, 2N903, SK3122.
2N5381	A5T3904, 2N904.
2N5382	A5T3905, 2N905, SK3114.
2N5383	A5T3906, 2N906, SK3114.
2N5386	2N5380.
2N5389	2N5388.
2N5391	E231.
2N5392	E305.
2N5393	E305.
2N5394 to 8	E300.
2N5399	MPS706, SK3039.
2N5400/1	BF398.
2N5404 to 7	2N5384.
2N5413	MPS6530.
2N5414	2N4409.
2N5415	2N5401.
2N5417	2N4123.
2N5425	MPS6566.
2N5432/3	E108.
2N5434	E109.
2N5447	BC177-307/A-328, MPS3638A.
2N5448	BC307/VI, MPS3638, SK3114.
2N5449	BC327/40, MPS6514, SK3024.
2N5450	BC327/16, BF178-297, MPS3705, SK3024.
2N5451	BC327/16, BF178-298, MPS6514, SK3024.
2N5452/3/4	E202.
2N5455/6	MPS3638A.

2N5457	E305
2N5458/9	E304, SK3112/3116.
2N5460 to 5	E176.
2N5471 to 3	E176.
2N5474/5	E175.
2N5476	E174.
2N5484/5	E305.
2N5486	E304.
2N5505 to 8	E176.
2N5509	E175.
2N5510 to 3	E176.
2N5514	E175.
2N5515/6	E400.
2N5517	E401.
2N5518	E402.
2N5519	E402.
2N5520/1	E400.
2N5522	E401.
2N5523/4	E202.
2N5543/4	E402.
2N5545	E400.
2N5546	E401.
2N5547	E402.
2N5548	E270.
2N5549	E113.
2N5550/1	BF176-178
2N5555	E114.
2N5556	E231.
2N5557	E232.
2N5558	E231.
2N5561 to 3	E231.
2N5564/5	E420.
2N5566	E421.
2N5592 to 4	E305.
2N5638	E111.
2N5639	E112.
2N5640	E113.
2N5647	E201.
2N5648/9	E202.
2N5650 to 2	MPS6568, SK117.
2N5653	E112.
2N5654	E113.
2N5655	BF338, SK3103.
2N5668	E305.
2N5669/0	E304.
2N5716	E201.
2N5717	E202.
2N5718	E203.
2N5777	2N4124.
2N5778	2N4400.
2N5779	2N4124.
2N5780	2N4400.
2N5794 to 6	2N4400.

2N5810	BFR41, MPS3706.
2N5811	BFR81, MPS3706.
2N5812	BFR41, MPS3706.
2N5813	BFR81, MPS6522.
2N5814	BFR50, MPSA40.
2N5815	BFR60, MPSA70.
2N5816	BFR40, MPSA20.
2N5817	BFR80, MPSA70.
2N5818	BFR40, MPSA20.
2N5819	BFR80, MPSA70.
2N5820	MPSA05, SK3024.
2N5821	MPSA55, SK3025.
2N5822	MPSA05.
2N5823	MPSA55.
2N5824 to 6	MPSA20, TIS92.
2N5827/8	MPSA20.
2N5829	MPS6535.
2N5835	2N4265.
2N5845	TIS116.
2N5901	MPSA18.
2N5902/3	E400, MPSA18.
2N5904	E401.
2N5905	E402.
2N5906/7	E400.
2N5908	E401.
2N5909	E402.
2N5911	E420.
2N5912	E421.
2N5949	BC337, E304.
2N5950	E304.
2N5951	E304.
2N5952/3	E305.
2N5964	MPSA42.
2N5965	MPSA43.
2N5998	MPSA18.
2N5999	2N5086.
2N6000	MPS6531.
2N6001	MPS6534.
2N6002	MPS6531.
2N6003	MPS6534.
2N6005	2N6067.
2N6006	2N5845.
2N6007	2N6067.
2N6008	MPSA18.
2N6009	2N5086.
2N6014	2N5845A.
2N6015	2N6067.
2N6016	2N5845A.
2N6017	2N6067.
2N6098	BDX70.
2N6099	BDX71.
2N6100	BDX72.
2N6101	BDX73.
2N6102	BDX74.

2N6103	BDX75
2NU74	AD133-133III-133IV, ADZ11, CTP1504-1508.
2S31	AF132-137-185, SK3005.
2S32	AC126-131, 2N2431, SK3004.
2S33	AC131-132-153, 2N610-2431, SK3004.
2S34	AC131, 2N2431, SK3004.
2S35	AF133-137-185, SK3008.
2S36	AF127-185, ASY26, 2N112, SK3008.
2S37	AC131, 2N2431, SK3004.
2S38	AC131-153, 2N2431, SK3004.
2S39/40	AC128, 2N2431, SK3004.
2S41	AF135, 2N2836, SK3009.
2S41A	AF134-135, AD150, 2N269-2836, SK3009.
2S42	AUY28, SK3009.
2S43	2N2431, SK3004.
2S44	AC131, SK3004.
2S45	AF133-137, SK3005.
2S49	AF133-137-185, SK3005.
2S52	AF185, SK3005.
2S53	AF133-137-185, 2N409-410, SK3005.
2S54	AC131-153, SK3004.
2S56	AC131, 2N2431, SK3004.
2S60	AF185, SK3005.
2S60A	AF185, 2N412.
2S91	AF185, SK3005.
2S92/A	AF185, 2N411, SK3008.
2S93	AF185, 2N140, SK3008.
2S93A	AF185, 2N412, SK3008.
2S101	2N726, SK3122.
2S103	AFY16, BF272, BFY11, BSW19-72, 2N2273-2873-3324-5354, SK3024.
2S104	AFY16, BF272, BFY19, BSW19-72, 2N2273-2873-3324-5354, SK3024.
2S109	AF135-185, SK3008.
2S110	AF115-135-136-185, SK3007.
2S112	AF135-136-185, SK3008.
2S141	AF115-134-136-185, 2N371-372, SK3006.
2S142	AF115-135-136-185, 2N370, SK3006.
2S143	AF115-135-185, SK3006.
2S144	AF117-137-185.
2S145	AF115-116-134-136-185, 2N1110, SK3006.
2S146	AF136-185, SK3006.
2S148	AF136-185, OC44.
2S155	AF185, SK3005.
2S159	AC162-170-185, SK3005.
2S160	AF185, SK3005.
2S163	AC131-153, OC72, 2N2431, SK3004.
2S178	AF185, 2N140, SK3005.
2S179	2N217-2431, SK3004.
2S303	SK3025.
2S321	OC203, SK3114.
2S322	OC200, SK3114.
2S322A	OC204, SK3114.

2S323	OC201, SK3114.
2S371	BFY10, 2S102.
2SA12	AF126-131-132-137-185-196, AFY15, ASY27, 2N412-642-4034, 2SA240, SK3005.
2SA13	AC131-137-185, 2N642-4034, 2SA240, SK3005.
2SA14	AF132-137-185, SK3005.
2SA15	AF131-132-136-185-196, AFY15, 2N641, 2SA240, SK3005.
2SA16	AF126-127-133-137-185, 2N642-4034, 2SA240, SK3005.
2SA17	AC122-125-151, AF134-136-185, OC304, 2N412-486, SK3005.
2SA18	AF124-130-131-134-136-185-194, 2N412-579, SK3005.
2SA19 to 21	AF185, SK3008.
2SA22	AF132-136-185, 2N411.
2SA23	AF126-133-137-185, 2N642-4034, 2SA240.
2SA24	AF178-194, 2SA235, 2N346-2495-4035.
2SA25	AF131-178-194, 2SA235, 2N346-2495-4035.
2SA27	AF125-134-135-185, 2SA156, 2N1110.
2SA28	AF127-134-137-185, 2SA240, 2N642-4034, SK3005.
2SA29	AF125-134-185, SK3008.
2SA30	AF132-136-185-196, AFY15, 2SA240, 2N641, SK3005.
2SA31	AF126-132-133-136-137-185-196, AFY15, 2SA240, 2N374-409-646-4034, SK3005.
2SA35	AF126-127-132-136-137-185-196, AFY15, 2SA240, 2N411-412-646-4034, SK3005.
2SA36	AF126-132-136-185-196, 2N411-412, SK3005.
2SA37	AF117-126-137-185, 2SA240, 2N646-4034, SK3005.
2SA38	AF126-127-134-136-137-185, 2SA240, 2N646-4034, SK3005.
2SA39	AF127-134-137-185, 2SA240, 2N646-4034, SK3005.
2SA40	AF126-132-136-185-196, SK3005.
2SA41	AC150, AF126-127-132-134-136-137-196, 2SA240, 2N646-4034, SK3005.
2SA42	AF126-131-132-134-136-137-196.
2SA43	AF131-135-185, 2SA156, 2N1110, SK3007.
2SA44	AF185, SK3005.
2SA44B	AF126-132-136.
2SA45	AF185, SK3006.
2SA48	AF131-135-185, 2N1110, 2SA156.
2SA49	AF132-135-136-185-196, AFY15, SK3005.
2SA50	AC124R-128-152IV, AF185, 2S32, 2N610-4106, SK3005.
2SA51	AF136-137-185, 2SA240, 2N642-4034, SK3005.
2SA52	AF126-131-134-137-185, 2SA240, 2N486-641-1058, SK3005.
2SA53	AF137-185, 2SA240, 2N614-642-4034, SK3005.
2SA57	AF185-194, 2SA235, 2N307-346-4035, SK3008.
2SA58	AF120-127-130-185-194, 2SA235, 2N346-4035, SK3006.
2SA59	AF137-185, SK3007.
2SA60	AF130-134-136-185-194, 2SA235, 2N346-393-4035, SK3006.

2SA61	AF185, SK3005.
2SA64	AF126-132-136-196, SK3008.
2SA65	2N36, SK3005.
2SA66	ASY27, 2SA65, 2N36-302-1307, SK3005.
2SA67	ASY27, 2SA65, 2N36-302-1309, SK3006.
2SA69	AF178, 2N2495, SK3006.
2SA70	AF118-178-186, AFY15, 2SA76, 2N370-2495-2707, SK3006.
2SA71	AF131-178-186-194, 2SA235, 2N346-2495-4035, SK3006.
2SA72	AF126-127-132-134-136-137-185-196, 2SA240, 2N642-4035, SK3006.
2SA73	AF127-133-134-137-185-197, 2SA240, 2N642-4035, SK3007.
2SA74	AF124-131-134-185-194, 2SA235, 2N346-4035, SK3006.
2SA75	AF126-127-133-134-137-185-197, 2SA240, 2N642-4034, SK3007.
2SA76	AF126-134-136-137-185, 2SA240, 2N641, SK3006.
2SA77	AF121-124-126-131-134-137-185, 2SA240, 2N641, SK3008.
2SA78	ASZ20, SK3006.
2SA80	AF126-134-137-185, 2SA240, 2N641, SK3006.
2SA81	AF132-136-185, 2N1634-1638, SK3008.
2SA82	AF134-136-185, SK3008.
2SA83	AF134-136-185, 2N1634-1638, SK3007.
2SA84	AF132-136-185, 2N1634-1638, SK3007.
2SA85	AF126-131-136, 2N544.
2SA86	AF126-131-135-137, 2SA240, 2N641.
2SA89	AF131-134-136, 2N481.
2SA90	AF131-135-136, 2N499, SK3006.
2SA92	AF125-131-132-135-136-185-195, 2SA156, 2N1110, SK3007.
2SA93	AF132-134-136-185-196, 2SA240, 2N308-417-641, SK3007.
2SA94	AF185, SK3008.
2SA100	AF127-133-136-137, 2SA240, 2N642-4034, SK3005.
2SA101	AF133-137-185-197, 2SA240, 2N642-4034, SK3007.
2SA101A	AF127, SK3007.
2SA102	AF124-125-127-133-137-185-197, 2SA240, 2N642-4034, SK3008.
2SA103	AF124-125-126-127-133-137-181-185-197, 2SA240, 2N642-4034, SK3006.
2SA104	AF127-133-135-137-185-197, 2SA240, 2N642-4034, SK3008.
2SA105	AF178, 2N299-2495, SK3008.
2SA106/7	AF185, SK3007.
2SA108	AF125-135-185, 2SA156, 2N374-1110, SK3007. .
2SA109	AF127-136-137-185, 2SA240, 2N374-481-642-4034, SK3007.
2SA110	AF126-131-136-185, 2SA240, 2N273-374-641, SK3007.

2SA111	AF126-135-136-185, 2SA240, 2N373-374-641, SK3007.
2SA112	AF125-135-136-185, 2SA156, 2N373-374-1110, SK3007.
2SA113/4/5	AF185, SK3007.
2SA116	AF135-178-194, 2SA235, 2N346-2495-4035, SK3006.
2SA117	AF124-130-134-135-194, 2SA235, 2N346-1178-2495-4035, SK3006.
2SA118	AF124-130-134-136-194, 2SA235, 2N346-1197-2495-4035, SK3006.
2SA121	AF135-178-185, 2SA156, 2N1110, SK3006.
2SA122	AF178-185-212, 2T201, 2N1178-1180, SK3006.
2SA123	AF125-135-178-185, AFZ12, 2T201, 2SA156, 2N1110-1179-1180, SK3006.
2SA124	AF124-134-135-194, AFZ12, 2T203, 2SA235, 2N346-1180-2495-4035, SK3006.
2SA125	AF124-134-194, AFZ12, 2T205A, 2SA235, 2N346-1180-2495-4035, SK3006.
2SA128/9	AF185, SK3008.
2SA130/1/2	AF185, SK3005.
2SA133	AF185, SK3006.
2SA134	2N2495, SK3006.
2SA135	2N2495, SK3005.
2SA136	AF126-132-136-185-196, AFY15, 2SA52, SK3006.
2SA137	AF185, SK3005.
2SA138	AF185.
2SA139	AF185, SK3006.
2SA141	AF126-127-132-133-136-137-185-196, AFY15, AFY27, 2SA55, 2N799, SK3006.
2SA142	AF126-127-132-133-136-137-185-196, AFY15, ASY27, 2N135- 799, SK3008.
2SA143	AF131-136-185, SK3006.
2SA144	AF124-130-134-185-195, 2SA235, 2N346-624-4035, SK3005.
2SA145	AF130-134-185, SK3005.
2SA146	AF133-136-185, 2N409, SK3005.
2SA147	AF133-137-185, SK3005.
2SA148/9	AF132-137-185, SK3005.
2SA151	AF185, SK3005.
2SA152	AF132-136-185, SK3005.
2SA153	AF134-178-185, 2N2495, SK3006.
2SA154	AF185, SK3006.
2SA155	AF134-137-185, 2N267, SK3006.
2SA156	AF135-185, SK3006.
2SA157	AF134-178-185, 2N2495, SK3006.
2SA159	AF134-185, 2N2495, SK3006.
2SA160	AF185, SK3005.
2SA161	AF106-178-196, 2SA230, 2N2495, SK3006.
2SA167	AF133-136-105, SK3005.
2SA168	AF132-137-185, 2N135, SK3005.
2SA168A	AF116-127-185, 2N397, SK3005.
2SA175	AF178-185, 2N2495, SK3006.
2SA176	AF185, SK3008.

2SA178	AF131-134-185.
2SA180	AF131-136-185, SK3005.
2SA181	AF133-137-138, 2N254, SK3005.
2SA182	AF133-137-185, SK3005.
2SA183	AC121, AF106-126-132-136-137-185-190-196, 2SA230, SK3005.
2SA184	AF133-137-185, 2N292.
2SA186	AF126-132-136-196, AFY15, 2SA52.
2SA188	AF185, SK3005.
2SA189	AF185, 2N410, SK3005.
2SA190/1	AF185.
2SA192	AF131-136-185, 2N411.
2SA193	AF133-137-185, 2N313.
2SA194	AF185.
2SA195	AF131-137-185, 2N135.
2SA196	AF127-133-185, 2N131.
2SA197	AF132-137-185, SK3004.
2SA198	AF131-137-185, SK3005.
2SA199	AF132-137-185.
2SA200	AF131-137-185.
2SA201	AC121, AF126-132-136-185-196, AFY15, 2SA52, SK3005.
2SA202	AC121V, AF127-133-137-185-197, AFY15, 2SA52, SK3005
2SA203	AF126-127-133-137-185-197, AFY15, 2SA52, SK3005.
2SA206	AF133-137-185, SK3005.
2SA208	AF126-130-136, ASY26, 2SA155, 2N799, SK3005.
2SA209	AF127-133-137, ASY26, 2SA155, 2N412-799, SK3005.
2SA210	ASY26, 2SA155, 2N799-1307, SK3005.
2SA211/2	ASY26, 2SA155, 2N799, SK3005.
2SA213	AF129-178, 2N2495, SK3006.
2SA214	AF130-135-178, 2N2495, SK3006.
2SA215	AF135-185, SK3006.
2SA216	AF178, 2N2495, SK3006.
2SA218	AF126-134-185, SK3008.
2SA219	AF185, SK3006.
2SA220	AF134-136-185, SK3008.
2SA221	AF126-134-185, SK3008.
2SA222	AF134-136-185, SK3008.
2SA223	AF126-134-185, SK3006.
2SA224	AF134-135-185, SK3008.
2SA226	AF134-135-185.
2SA227	AF178, 2N2495, SK3006.
2SA229	AF102-139-178, M9031, 11M139, GMO290, 2N2495 SK3006.
2SA230	AF102-139-186, GM0290, M9031, 11M139, 2N2495, SK3006.
2SA233	AF117-126-185, 2N136, SK3007.
2SA234/5	AF106-178-185-186-190, GM0790, 2SA230, SK3006.
2SA236	AF185, SK3007.
2SA237	AF185, SK3008.
2SA239	AF106-190, GM0790, 2SA230, SK3006.

2SA240	AF106-185-190, GM0790, 2SA230,2N2495, SK3006.
2SA241	AF102-178, 2SA76, 2N2207-2495, SK3006.
2SA242	AF118, 2N2495.
2SA243	AF118-178, 2N2495.
2SA246	AF106, GM0790, SK3006.
2SA250	AF185, SK3008.
2SA253	AF185, SK3006.
2SA254	AF136-185-186, 2N1058, SK3005.
2SA255	AC133-137-152-162-170, 2SB459, 2N37, SK3005.
2SA256	AF135-185, SK3008.
2SA257	AF130-185, SK3007.
2SA258	AF130-134-185, SK3007.
2SA259	AF185, SK3008.
2SA266	AF130-134-135-185-194, 2SA235, 2N346-4035, SK3008.
2SA267	AF125-131-135-185-195, 2SA156, 2N1110, SK3006.
2SA268	AF126-132-136-185-196, SE316, 2SA240, 2N641, SK3006.
2SA269	AF127-133-185-197, 2SA240, 2N642-4034, SK3008.
2SA270	AF125-131-133-137-185-195, 2SA156, 2N1110, SK3007.
2SA271 to 74	AF127-132-133-137-185-197, 2SA240, 2N642-4034, SK3008.
2SA275	AF125-135-185-195, 2SA156, 2N1110, SK3008.
2SA276	ASZ21.
2SA282	ASZ21, SK3005.
2SA285/6/7	AF132-185, SK3008.
2SA288	SK3008.
2SA289/90	SK3006.
2SA293	AF126-132-136-137, SET316, 2SA240, 2N267-641, SK3006.
2SA296/7	AF126-132-136-196, AFY15, SET316, 2SA52, 2N641, SK3005.
2SA298	AF185, SK3008.
2SA307	AF185, SK3008.
2SA311/2/3	AF185, SK3008.
2SA314	AF125-134-185, SK3008.
2SA315/6	AF134-135-185, 2SA156, 2N1110, SK3008.
2SA321	AF127-133-137-197, 2SA240, 2N642-4034, SK3006.
2SA322	AF185, SK3008.
2SA323	AF124-134-185-194, 2SA235, 2N346-4035, SK3008.
2SA324	AF125-131-135-185-195, 2SA156, 2N1110, SK3006.
2SA340	AF135-185, 2SA156, 2N1110.
2SA341	AF121-130-185-194, 2SA235, 2N346-4035, SK3008.
2SA342	AF121-130-134-178-185-194-195, 2SA235, 2N346-2495-4035, SK3006.
2SA343	AFZ11, SK3006.
2SA350	AF125-126-127-130-132-135-137-196, SET316, 2SA240, 2N641, SK3006.
2SA352/3	AF124-126-129-130-134-137, SET316, 2SA240, 2N641, SK3006.
2SA354/5	AF139, M9031, 11N139, 2N2244, SK3006.

2SA356	AF139, M9031, 11N139, 2N2244, SK3008.
2SA361	SK3007.
2SA368	2SA468, SK3007.
2SA369	2SA469, SK3008.
2SA370	2SA470.
2SA371	AF127-137, 2SA240, 2N642-4034.
2SA372	2SA472.
2SA374	2SA472.
2SA377	AF106-190, 2SA230, GM0760, SK3006.
2SA378	2SA478.
2SA379	2SA479.
2SA380/1	AF185, SK3007.
2SA382/3/4	AF185, SK3008.
2SA385	2SA485, SK3006.
2SA386	2SA486.
2SA395	2SA495, SK3006.
2SA396	2SA496.
2SA397	2SA497.
2SA398	2SA498, SK3006.
2SA399	2SA499, SK3006.
2SA400	AF185, SK3008.
2SA420/2	AF139, GM0290, M9031, 11N139, 2N2244, SK3006.
2SA433	AF126-132-136-137-196, SET316, 2SA240, 2N641, SK3006.
2SA454/5/6	AF139, SK3006.
2SA471	AF126-127-132-136, SK3006.
2SA495	2N4060.
2SA561	BC212L, SK3025.
2SA562	2N3702, SK3025.
2SA640	2N5086.
2SA1018	AF185.
2SB17	AD148-149, OC30, 2N351.
2SB22	AC128-131-132-152-152V-184, OC318, 2SB156, 2N238-1924, SK3004.
2SB25	AD131-138/50-139-150, ASZ18, AUY22, 2N297A, SK3009.
2SB26	AD130-138-139-148-150, 2SB151, 2N561-1022A-2836-3617, SK3009.
2SB27	AD148-150, 2N376-2836, SK3009.
2SB28/9	AD148-150, 2N351-2836, SK3009.
2SB30	AD148, OC30, 2N255-2836, SK3009.
2SB31	AD148-150, 2N256-2836, SK3009.
2SB32	AC122R-126-151-151IV, 2SB459, 2N238-680-2429, SK3004.
2SB33	AC122-125-128-131-151-153V, OC304, 2SB459, 2N238-651-2431, SK3004.
2SB34	AC122-125-126-131-151-173, 2SB459, 2N238-1381-2431, SK3003.
2SB37	AC122-125-126-128-151-173, OC318, 2SB459, 2N238-2431, SK3004.
2SB38	AC117-122-125-126-131-151-153-173, OC318, 2SB459, 2N238-2431, SK3004.
2SB39	AC122-125-126-151-151IV, OC304, 2SB459, 2N238-2429, SK3003.

2SB40	AC122/30-125-126-151, 2SB459, 2N238-524, SK3004.
2SB41	AD131-150, TI3029, 2SB471, 2N2065-2836, SK3009.
2SB42	AD132-149-150, ASZ15, 2SB472, 2N2065, SK3009.
2SB43	AC122-125-151, OC304, 2SB459, 2N238-651, SK3003.
2SB44	2N2431, SK3004.
2SB46	AC162-170, 2N2429, SK3003.
2SB47	AC122-125-132-151-152IV, OC304, 2N651-2429, SK3003.
2SB48	AC126, 2N2429, SK3004.
2SB49	AC128-131, 2N2429-2431, SK3004.
2SB50	AC153, 2N2429, SK3003.
2SB51	AC131, 2N2431, SK3003.
2SB52	AC131-153, 2T324, 2N1413-2431, SK3003.
2SB53	AC131-153, 2T383, 2N1307-2431, SK3004.
2SB54	AC151-151VI/VII-162, OC304, 2SB43, 2N408-2431, SK3004.
2SB55	AC131/30-151-152, 2SB459, 2N238-525-2431, SK3004,
2SB56	AC126-128-131-151V-152-184, OC318, 2N2431, SK3004.
2SB57	AC122-125-126-151-152V, OC304, 2N651-2431, SK3004.
2SB58	2N2431.
2SB59	AC122G-125-151VI, SK3004.
2SB60	2N2429, SK3004.
2SB60A	AC126-128-131-151V/VI, 2N2429, SK3004.
2SB61	AC117-122G-125-151VI-153, 2N680-2429, SK3004.
2SB62	AD139, SK3009.
2SB63	AD149, 2N2836, SK3004.
2SB65	AC122-125-126-128-131-151/V-152, OC304, 2N36-651-1394, SK3004.
2SB66	AC126-128-131-151/VII-173, OC318, 2N2429-2431, SK3004.
2SB67	AC124-128-153-180, 2SB222, 2N467-4106, SK3004.
2SB68	AC122-122/30-125-151, ASY48-81, 2SB460, 2N524, SK3004.
2SB70/1	2N2429, SK3004.
2SB73	AC122-125-126-151-152IV, OC304, 2SB459, 2N2429, SK3004.
2SB74	AC126, 2N2429, SK3004.
2SB75	AC122-125-151-162-163-170, OC304, 2N34-2429, SK3003.
2SB75A	AC122-151/V, 2SB459, SK3004.
2SB76	AC151, 2N2429, SK3003.
2SB77	AC125-131-151/VI-163-184, 2N41-2431, SK3004.
2SB78	AC128, 2N2431, SK3004.
2SB79	2N187A-2431, SK3004.
2SB80	AD139, 2N2836, SK3009.
2SB83	AD150, 2N2836, SK3009.
2SB84	AD150, SK3014.
2SB86	CTP1104.
2SB87	CDT1313, SK3004.
2SB89	AC122G-125-151VI-152IV, 2N2431, SK3003.

2SB90	AC126-162-170, 2N2429, SK3004.
2SB91	AC131-153, 2N2431, SK3004.
2SB92	2N2431, SK3003.
2SB94	AC131-151-173, OC304, 2SB459, 2N44A-651-2431-2447, SK3004.
2SB95	2N2431, SK3004.
2SB96	2N2431.
2SB97	AC126, 2N2429.
2SB98	AC125-153, ASY26, SK3004, 2N799-2431, 2SA155.
2SB99	AC152-153, 2N633-2431.
2SB100	AC126-151-162-173, ASY26, SK3004, 2N651-799-2429, 2SA155.
2SB101	AC128, ASY26, SK3004, 2N799-2431, 2SA155.
2SB102	2N2431.
2SB103	2N2431.
2SB104	AC152-153, 2N188A-2431.
2SB105	AC152, 2N241-2836.
2SB106	AD149, 2N2336.
2SB107	AD148, CTP1104, OC30, 2N353.
2SB107A	AD148-149-150-166, CTP1104, OC30, SK3009, TI156, 2N456-468, 2SB471.
2SB108/9	2N2836.
2SB110	AC126-151-162, SK3003, 2N2429.
2SB111	AC122-125-151VI-152-163, OC304, SK3003, 2N77-2429.
2SB112	AC125-131-151V-152IV, SK3004, 2N2429
2SB113	AC122-125-151/V-163, OC304, SK3004, 2N96-2429.
2SB114	AC122-125-151-163, OC304, SK3003, 2N36-2431.
2SB115	AC122-125-151/VII-163, OC304, SK3003, 2N37-2431.
2SB116	AC122-125-151/VII-163, OC304, SK3004, 2N38-2431.
2SB117	AC122-125-151/VII-163, OC304, SK3004, 2N38A-2431.
2SB118/9	2N2836.
2SB120	AC122Y-125-126-151V-163, SK3004, 2N951-2429.
2SB121	AC125-2N398.
2SB122/3/4	AS215, AUY22-28.
2SB125	AUY21.
2SB127	AD130-133-138, ADZ11, SK3009, 2N561-1022A, 2SB151.
2SB128	ASZ18, AUY22-28.
2SB131	2N2836.
2SB134	AC126, 2N2429.
2SB135	AC122Y-125-131-151V-153-163, SK3004, 2N138-2429.
2SB136	AC122Y-125-131-151VII-153, SK3004, 2N2431.
2SB137	2N2836.
2SB140	ASZ16, AUY21, 2N2836.
2SB141	ASZ15, AUY22-28, 2N301A, 2T3021.
2SB142	AD149-150, 2N2836.
2SB143	AD139-150, 2N669-2836, 2T3032.
2SB144	AD139-150, 2N669-2836, 2T3032.
2SB145	AD150, ST3042.
2SB146	AD150, ST3043.

2SB148	ASZ16, AUY21.
2SB150	2N398.
2SB151	AD138-30-138/50, ASZ18, AUY22, CDT1313, SK3009.
2SB152	2N157A.
2SB153	AC126-162, 2N2429.
2SB154	AC131-153, 2N2431.
2SB155	AC117-153-180, SK3003, 2N249-2431, 2SB370A.
2SB156	AC131-180, SK3004, 2N2431.
2SB156A	AC117-121IV-128-131-153, SK3004, 2N1307-2431.
2SB157	AC126, 2N2429.
2SB158	AC126, OC59, 2N2429.
2SB159	AC126, 2N2429.
2SB160	OC60, 2N2429.
2SB161	AC128-151, SK3004, 2N2431.
2SB162	AC122-125-151-162, OC304, SK3004, 2N2431.
2SB163	AC124-153, 2N320-2431.
2SB164	AC124-153, 2N321-2431.
2SB165	AC125-153, 2N323-2431.
2SB166	AC124-153, 2N2431.
2SB167	AC122-152V, SK3004.
2SB168	AC122-125-126-151/V, OC304, SK3003, 2N2429.
2SB169	AC122-126-128-151/VI, OC304, SK3003, 2N2431.
2SB170	AC126-128-151-162, SK3004, 2N76-2431.
2SB171	AC122-125-126-151/V-162-163.
2SB172	AC122-126-151IV, SK3004, 2N2431.
2SB173	AC122-125-126-151/V-162, OC304, SK3004, 2N2429.
2SB174	2N2431.
2SB175	AC125-126-128-151-151/VI, SK3004, 2N2431.
2SB176	AC125-151-175, OC318, SK3004, 2N2431-4106-4607, 2SB222.
2SB178	AC124R-152V, 2N610-2431, 2SB222.
2SB178A	ASY80.
2SB179	AC128, 2N2431.
2SB180	2N2836.
2SB181	AD148-149, 2N307.
2SB182	AF185.
2SB183	AC126, 2N132A-2429.
2SB183A	OC60.
2SB184	AC125-163, 2N63-2431.
2SB185	AC122-125-128-151V-163, OC304, SK3004, 2N107.
2SB186	AC122-125-128-151-152VI-163, OC305, SK3004, 2N610-2429-4106, 2S32.
2SB187	AC122-126-128-151-152VI, OC304, SK3004, 2N610-4106, 2S32.
2SB188	AC117R-122-125-151-152-152VI, OC304, SK3004, 2N610-2431, 2S32.
2SB189	AC131-152VI-180, SK3004, 2N610-2431, 2S32.
2SB190	AC125-163, 2N322, 2N2431.
2SB191	AC122-125-163, 2N47.
2SB192	AC125-163, 2N48-2431.
2SB193	2N2431.
2SB194	AC128-2N2431.
2SB195	AC117-153, 2N1097-2431.

2SB196	2N2431.
2SB197	AC124-153, 2N226-2431.
2SB198	AC124-153,
2SB199	AC124-153,
2SB200	AC117-124-128-153V-153/VI-180, SK3004, 2N467-2431-4106, 2SB222.
2SB201	AC117-128-153V-162, 2N467-4106, 2SB222.
2SB202	AC117-139-149-153/VII-180, SK3004, 2N467-4106, 2SB222.
2SB215	AD163, ASZ15, SK3009, 2SB341.
2SB216	AD131-138/50-149, SK3034, TI3029, 2N2065, 2SB471.
2SB217	AD150, SK3009, TI156, 2N456.
2SB219	AC153-163, 2N41-2431.
2SB220/1	2N2431.
2SB222	AC124-153, 2N220-2431.
2SB223	2N2431.
2SB225	AC117-128-153, 2N43.
2SB227	AC124-128-153, 2N223-527.
2SB231	AU103-105, MP939, 2N1906.
2SB240	CTP1104.
2SB240A	AD148, CTP1104, OC30, 2N141.
2SB242	AD148, ASZ15, AUY18, CTP1104, OC30, SK3009, 2N1183.
2SB242A	AD148, CTP1104, OC30.
2SB247	AD133IV-133V-138, ADZ11, SK3009.
2SB248A	AC105-117-153, AD133IV-133V-138, ADZ11, SK3009, 2N1501.
2SB250	AD133V-138, ADZ11, SK3009.
2SB250A	AC117-128-153, 2N1245.
2SB251	AD133V-138, ADZ11, SK3035.
2SB252	AD131-133V-138/50, ADZ11, SK3035, 2N540-1551.
2SB253	AC128-153, ASZ218, 2N1263.
2SB254	AC128, 2N2431.
2SB255	2N2431.
2SB257	AC150-151R-161, OC306.
2SB261/2	AC126, 2N2429.
2SB263	2N2431.
2SB264	AC125-126-153, 2N133-2429.
2SB266	AC125-152-170.
2SB267	AC128-131-153.
2SB290	AC122-125-151-161, ASY26, OC304, SK3005, 2N799. 2SA155.
2SB291	AC131, ASY26, SK3004, 2N799, 2SA155.
2SB292	AC125-131-151-173, ASY26, OC318, SK3004, 2N799, 2SA155.
2SB302	AF124-134, SK3004, 2SB235.
2SB303	AC125-150-151R-161-163, OC306, SK3004, 2SB73.
2SB345	AC122V-125-131/30-132-151VII-152-173, SK3004, 2N525-2429.
2SB346	AC131/50-151VII, 2N2429.
2SB347	2N2429.
2SB348	AC126, 2N2429.

2SB364	AC125-128-131-151-153-173, OC318, SK3004, 2N1924-2431, 2SB370A.
2SB365	AC128-131-153, SK3004, 2N1924-2431.
2SB370	AC128-131/30-132-152-184, OC318, SK3004, 2N525.
2SB371	AC122G-125-151VI, SK3004.
2SB378	AC121-125-128-131-151-173, OC318, SK3004, 2N2431, 2SB156A.
2SB379	AC121-128-131, SK3004, 2N2431, 2SB156A.
2SB381	AC122R-126-151IV, SK3004, 2N238, 2SB459.
2SB383	AC124-128-128K-153-180, SK3004, 2SB67.
2SB391	ASZ18, AUY21-28, SK3009, 2N352, 2SB339.
2SB400	AC116-121/VI-128-160-184, SK3004.
2SB419	AD148-152.
2SB422	AC122-126-151.
2SB439	AC128-163-171-173, SK3004, 2N506-2907, 2SB459.
2SB440	AC128-163-171-173, SK3004, 2N506-2907, 2SB459.
2SB443/A/B	AC121VI, SK3004.
2SB444/A/B	AC121VI, SK3006.
2SB473	AD155-162-162VII, SK3052, 2SB367.
2SB475	AC121VI, SK3004.
2SC11	ASY73.
2SC15	2N1753/4.
2SC23C	BD137.
2SC31	BC140, 2N2218/A-3020-3036.
2SC32	BFY13.
2SC32A	BFX55.
2SC33	BFX55-63.
2SC38	BFX55.
2SC41	BU110.
2SC43	BU110-2N1490.
2SC44	BU110.
2SC45	BFX55.
2SC50	BU110.
2SC53	BFX55.
2SC64	BSX22-45, 2N3252-4231.
2SC73	2N168A-1086, 2T73R.
2SC74	BFX55.
2SC75	2N169-293, 2T75R.
2SC76	2N169-293, 2T76R.
2SC77	2N169-293, 2T77R.
2SC78	2T78R.
2SC79	BF173, 2SC464.
2SC89	ASY74.
2SC90	2N1304.
2SC91	2N1306.
2SC97	BC140, 2N2218/A-3020-3036.
2SC103	BDY91.
2SC105	BC169/B-173-184B, MPS6521.
2SC120/1	BC140/VI, BSY46, 2N2218A-3036, 2SC708.
2SC122	BC140/Y, BSY46, 2N2218A-3036, 2SC708.
2SC123/4	BC140/XVI, BSY46, 2SC708, 2N2218A-3036.
2SC127	BC140/XVI, BSY46, 2SC708, 2N2218A-3036.
2SC147	2N2196.
2SC170	BC108-183, 2SC458, 2N3391.

2SC174	BC167-171A-183, MPS6566, 2SC458.
2SC191/2	2N1074/5.
2SC193	2N336/7.
2SC194	2N1074/5.
2SC195/6	2N1278/9.
2SC197	2N1277/8.
2SC199	BSX61-73, BSY34, 2SC458, 2N2218.
2SC200/1/2	BFX55.
2SC206	BF184-234, 2SC460.
2SC210/1	BSX61-73, BSY34, 2SC479, 2N2218.
2SC212	BSX22-45, 2N3252-4231.
2SC230	BSX49, 2N2222A.
2SC250	BFY12.
2SC281	BC107A-182A, 2SC458, 2N2921-3568.
2SC282	BC108/A-183/A, 2SC458, MPS6565, 2N3391.
2SC283	BC107B, BCY65, 2SC648, 2N2483/4.
2SC284	BSX49, 2N2222A.
2SC291	BSW65, BSX63; 2N1893.
2SC300/1	BSX48, 2N2222A.
2SC302	BSX49.
2SC313	BC108A-183A, 2SC458, 2N3391.
2SC316	BC109-184B, MPS6521, 2SC458.
2SC318	BC107A-182A, 2SC458, 2N2921.
2SC319	BSX48, 2N2221.
2SC350	BC107B-182B, 2N3568.
2SC366G	BC182L.
2SC367	2N3705.
2SC367G	BC183L.
2SC368	BC107B-182B, 2N3568.
2SC369/G	2N3711.
2SC370/1/2	BC168A-172A-183A, MPS6520, 2SC458, 2N3709/10.
2SC371T	BC182L.
2SC373	BC168A/B/C-172A-183A, MPS6520, 2SC458, 2N3711.
2SC374	BC168C-172C-183C.
2SC377/8	BC167A-171A-182A, MPS6566, 2SC458, 2N3709.
2SC379	BC169/B-173B-184B, MPS6521.
2SC380	BC167/A-172A-182A, MPS6566, 2SC458, 2N3709.
2SC381	BC168/A-173A-183A, BF224J, MPS6520, 2SC458.
2SC382	BF173-224/5J, 2SC464.
2SC382G/R	BF232.
2SC383	BF224J.
2SC384/5	BF173-224-232, 2SC464.
2SC386	BF173-224/J, 2SC464.
2SC387	BFX59, TIS18, 2SC707.
2SC388/9	BF173-224, 2SC464.
2SC394	BC167-171A-182A, MPS6566, 2SC458, 2N3709.
2SC398/9	BF125-185-237, 2SC535, 2N3709.
2SC401	BC148B-167B-171A-172-182A-183B, MPS6565/6, 2SC458.
2SC402/3/4	BC167B-171A/B-182A/B-147B, MPS6566, 2SC458.
2SC429	BF173-224, 2SC464.
2SC454	BC148/A-172A-183A, MPS6520, 2SC281, 2N3568.
2SC456	BSX73 BSY58, 2SC479, 2N2218.

2SC458-60	BC148/A-172A-183A, MPS6520, 2SC281, 2N3568.
2SC461	BC148/A-172A-183A, MPS6520, 2SC281, 2N3568.
2SC464	BF173-224.
2SC465/6	BF173-224, 2SC464.
2SC470	BF140D-178, 2SC856.
2SC475/6	BC179-206-253-259C, MPS6516.
2SC478	BC107-182A, 2SC458, 2N2921-3568.
2SC535	BF184-194-229-232-234-254.
2SC538	BC108/B-183A, 2SC458, 2N3391.
2SC561	BC108-183A, 2SC458, 2N3391.
2SC595	BSX48, 2SC495, 2N2221.
2SC601	BFX59, 2SC707.
2SC605	BF173-224, 2SC464.
2SC611	BFX59, 2SC707.
2SC622	BC108A/B-183A, 2SC458, 2N3391.
2SC631	BC107B-182B.
2SC632	BC108C-183C.
2SC633/4	BC107A-182A.
2SC640	BC108C-183C.
2SC644	BC167B-171B-182B, MPS6566.
2SC649	BC107A-182A, 2SC458, 2N2921.
2SC650	BC107B-182B.
2SC657/8/9	BF173-224, 2SC464.
2SC682/3	BF173-224, 2SC464.
2SC689A	BC107A-182A, 2SC458, 2N2921.
2SC689H	BC107A.
2SC712	BC168A-173, 2SC712A.
2SC732	2N3711.
2SC733	2N3710.
2SC734	BC107B-182B/L, 2N3508.
2SC735	2N3705.
2SC752G	TIS48.
2SC780G	BSS34.
2SC784	BF594.
2SC785	BF595.
2SC838/9	BC167-171A-182A, MPS6566, 2SC458.
2SC856	BF178.
2SC864	BC167-171A-182A, MPS6566, 2SC458.
2SC894	BC108/A-182A.
2SC912	BC108/A-182A.
2SC917	BF177.
2SC941	BC168/A-172A-183A, MPS6520.
2SC943	BC107-182A.
2SC1000	MPSA18.
2SC1010	BC110-174, 2SC856.
2SD30	AC179, 2SD96, 2N2430.
2SD31/2	AC127-179, 2SD96, 2N2430.
2SD33	AC127-179-181-186, 2SD96, 2N2430.
2SD34	AC179, 2SD06, 2N2430.
2SD35/6	AC127-179, 2SD96, 2N2430.
2SD37	AC179-181-186, 2SD96, 2N2430.
2SD38	AC127-175-179-181-187, 2SD96.
2SD43	AC127-185-186.

2SD43A/4	AC127-179, 2SD96, 2N2430.
2SD45	BU110.
2SD46/7	BD130, 2N3055.
2SD48/9	BUY46.
2SD50/1/3/4	BD130, 2N3055.
2SD56	BU110.
2SD57/8	BUY43.
2SD59/60	BD130, 2N3055.
2SD61	AC127-179, 2SD96, 2N2430.
2SD62	AC179, 2SD96, 2N647-2430
2SD63	AC179, 2T69, 2SD96, 2N647-2430.
2SD64	AC179, 2T64R, 2SD96, 2N647-2430.
2SD65	AC179, 2T65R, 2SD96, 2N647-2430.
2SD66	AC179, 2T66R, 2SD96, 2N647-2430.
2SD70	BD107-109-124, BDY34, 2N3054.
2SD71	BUY46.
2SD73	BD130, 2N3055.
2SD74	BU110.
2SD75	AC176-185-186-187K, 2N2430.
2SD77	AC127, 2N2430.
2SD79	BUY46.
2SD80/1/2/3	BD130, 2N3055.
2SD84	BU110. ·
2SD96	AC175K-187K.
2SD102	BU110.
2SD104/5	AC179, 2T690, 2SD96, 2N647-2430.
2SD118/9	BU110.
2SD124/A/5A	BD130, 2N3055.
2SD126H	BU110.
2SD141/2	BUY43.
2SD143	BUY46.
2SD144/6	BD107-109-124, BDY34, 2SC830, 2N3054.
2SD147	BDY13.
2SD151	BD130, 2N3055.
2SD152	BU110.
2SD154	BDY13.
2SD172/3	BD130, 2N3055.
2SD174	BUY46.
2SD175/6	BD130, 2N3055.
2SD178	2N2430.
2SD180	BD130, 2N3055.
2SD182	BUY43.
2SD183/4/5	BUY46.
2SD186/7	AC179, 2N2430.
2SD189A	BU110.
2SD195	AC127-179, 2N2430.
2SD196/7	BUY55IV.
2SD198	BU111.
2SD226	BUY43.
2SD226A	BUY46.
2SD234	BUY46.
2SJ11	E202.
2SK11/2	E231.
2SK13	E202.

2SK17G	E202.
2SK17.Orange	E201.
2SK17R	E201.
2SK17Y	E202.
2SK19B/6	E304.
2SK19Y	E305.
2SK23	E305.
2SK32	E305.
2SK35	E211.
2ST12	E202.
2ST15/6	E271.
3G2	E202.
3SK20H	E202.
3SK21H	E202.
3SK22/3	E211.
3SK28	E211.
3NU72	OC30A, OD603, TF77-78/301.
3NU73	OC26, TF80.
3NU74	AD133V, CTP1504-1508, 2N1146A/7A.
4G2	E304.
4NU74	AD131III/IV, CTP1504.
5G2	E305.
5NU73	TF80/60.
5NU74	AD131V, CTP1504, 2N1146B/7B.
6NU74	AD132III/IV, ADZ12, CTP1500-1503.
7NU74	AD132V, CTP1500-1503, 2N1146C/7C.
16K1/2/3	BF225J.
40022	AD162, MP2060, NKT452, 2N1539-2138.
40050	AD138, ADZ11, AUY21-33, MP2061, NKT404, SFT265, 2N1540-2139.
40051	AD138, ADZ11, AUY21-33, MP2062,NKT404,SFT266, 2N1540-2140.
40232	MPS5172.
40235	MPS6568.
40236	MPS6507.
40237	MPS3563.
40238/9/40	MPS6569.
40243	MPS6507.
40245	MPS6511.
40250	BD131, BDX24.
40251	BDX13, BDY38.
40254	AD162, AUY21, MP2060, NKT452, 2N1539-2138.
40314	BC140X.
40319	BC160X.
40360	2N3019.
40361	BC141X.
40362	BC161X.
40408	2N2218A.
40100	BSW65.
40421	AD138/50, ADZ12, AUY22-32, MP2062, NKT403, SFT 267, 2N154-2140.
40439	AU106, DTG1010, 2N5325.
40440	AU107, DTG1110, MP3730, 2N5324.

40462	AD138, ADZ11, AUY21-33, MP2061, NKT404, SFT265, 2N1540 21 39;
40594	BFX34.
40636	BD183, BDX23.
A104	BC108A-168A-170C-208A-386, GI3710, MPS3710, 2N4135, 40450, SK3020.
A106	BC108A-168A-170B-386B, GI3710, MPS3710, 2N4134, 40450, SK3020.
A108	BC108A-169B-170C-208A-386, GI3711, MPS3711, 2N4135, 40450, SK3016.
A110	BC108C-109C-169C-172C-209C-386C, GI3711, MPS3711, 2N3711, 40399.
A111	BC109B-169B-172B-208B-386B, GI3710, MPS3710, 2N3710, 40397, SK3122
A115	BF185-235-273D-311-385 ·BFY39/1, GI3709, MPS3394, NKT10519, 2N3394, SK3020
A116	BF186-234-273C-311-384, BFY39/II, GI3710, MPS3710, NKT10519, 2N3710, 40233, SK3122.
A130	BC110-141/6-254, BF156-177, ME1120, 2N699-1990N 4410-5184, SK3039.
A132	BC110-141/6, BF156-177, ME1120, 2N699-1990-1990N-4410-5184, SK3017A.
A133	BC285, BFY45-65, BSY55, ME1100, 2N1573-1893, 2509-4410-5185, SK3045.
A157B	BC107B-207B-237B-272-385, BCY56B-59B, MPS6575, 2N3242A-3566, SK3122.
A157C	BC107C-207C-272-237C-386B, BCY56C-59C, MPS6575, 2N3242B.
A158C	BC108B-108C-208B-208C-238C-386A, BCY58C, GI3710, MPS3710, 2N3710, 40450, SK3020.
A159B	BC109B-209B-239B-386C, BCY58B, GI3710, MPS3710, 2N3710, 40397, SK3020.
A159C	BC109C-209C-239C-386C, BCY58C, GI3711, MPS3711, 2N3711, 40450, SK3122.
A170	BC177-204-212-261A-291D-307, 2N3245-3906-4121, 40406, SK3025.
A171	BC178-205-208-213-262A-291A, 2N3244-3906-4121, 40419.
A177	BC177A-212A-261B-291D-307A, ME0411, 2N3233-3964-4349, 40406.
A178A	BC178A-205A-213A-262B-291A-308A, 2N3964-4059-4359, 40419, SK3114.
A178B	BC178B-205B-224-263A-292A-308B, 2N3964-4058-4359, 40419.
A179A	BC179A-205B-224-263B-292A-309A, 2N3964-4059-4359, 40419, SK3114.
A179B	BC179B-179C-206B-224-263C-292B-309B, 2N3964-4059-4359, 40419, SK3114.
A192	E304.
A210	BFS50, BFW16A-47, BFX55, MM8002, 2N1479-3137-3866-3948-4875, 2SC890.
A211	BFS50, BFW17A-47, BFX55, MM8003, 2N1479-3137-3866-3948-4875, 2SC890, 40280.
A270	BFW47, BFY99, BSW28, 2N1483-3553-3866, 2SC890.
A274	BFW47, BFY99, BSW28, 2N1483-3553-3866, 2SC890.

A301	BCY66, BF176-225-240-241, MPS6575, SE1002, 2N930-2692, SK3122, 40236.
A306	BC170A, BF184-224-234-254-311, GI3392, SE2001, 2N2708-4124, SK3122.
A307	BC170B, BF184-185-224-235-255-311, GI3392, SE2002, 2N2708-4124, SK3122.
A310	BC145-254, BF177, BFY45-65, BSY55, ME1100, 2N699-1990-1990N-4410-5184, SK3045.
A311	BC110-141/6-142-254, BF177, ME1120, 2N699-1990-1990N-4410-5184, SK3024.
A321	BC108B-109B-132-168B-208B-238B, GI3707, MPS3707, SE4001, 2N3707, 40309, SK3039.
A322	BC108C-168C-208C-238C-GI3708, MPS3708, SE4002, 2N3708, 40311.
A323	BC108B-168B-208B-238B, GI3710, MPS3710, SE4003, 2N3710, 40315.
A324	BC168C-208C-238C, BCY58B, GI3711, MPS3711, SE4003, 2N3711, 40309.
A344	ASY17, BSX12, GI3393, ME3011, MM5224, SE3001, 2N708-743-914-2865-3478-3854, SK3122.
A345	BSX12, BSY18, GI3393, ME3001, MM5224, SE3001, 2N708-744-914-4252-5133-5179, SK3122.
A346	BSW59, BSX12, BSY62, GI3392, ME3002, MM5224, SE3001, 2N706A-708-4252-5133-5179, SK3122.
A415	BF167-173-184-185-195-224-235, GI3392, SE2001, 2N2708-4274, SK3018.
A417	BF167-184-194-224-234, GI3392, SE2002, 2N2708-4274, SK3039.
A418	BF167-184-194-224-234-254, GI3392, SE2001, 2N3932-4124, SK3039.
A419	BF167-185-195-225-235-255, GI3393, SE2002, 2N3932-4124, SK3039.
A420	BF167-185-195-225-235, GI3393, SE2002, 2N3932-4124, SK3039.
A427	BF183-305, BFW30, BFX59, BFY78-90, ME3002, MM1501, 2N918-3572-5179-5201, SK3039.
A430	BF183-305, BFX59, BFY78-90, ME3002, MM1500, SE3005, 2N918-3570-5179, SK3039.
A451	BF115-167-184-224-234-311, 2N2857-2921-4134-4874, SK3117.
A454	BF167-184-194-224-234-254, 2N2857-2921-4134-4874, SK3039.
A455	BF167-184-185-194-224-234-254, 2N2857-2921-4134-4874, SK3039.
A466	BF167-184-196-198-237-251, 2N3724-4427-4876-5144, SK3024.
A472	B161-173-199-310, BFX60, 2N4072-4137-4876-5236, 40240.
A473	BF161-173-196-199-225-310, BFX55, BFY99, MM1805, SE5006, 2N3866, SK3018.
A480	BF159-180-311-384, BFS92, BFX62, BFY99, MPS6507, SE3005, 2N918-4253, 40238, SK3039.
A481	BF159-181-311-384, BFS92, BFX59, BFY99, MPS6507, SE3005, SK3039, 2N918-4253, 40238.

A482	BF182-274-311-384, BFS92, BFX62, BFY99, MPS6546, SE3005, 2N918-4253, 40238, SK3117.
A483	BF183-274-324, BFS92, BFX62, ME3011, 2N918-.-2616-2708-2729-3563, SK3117.
A484	BF161-173-199-200-385, BFX55, MPS6548, SK3117, 2N2616-2729-3600-4137.
A485	BF183, BFX89, BFY78-90, ME3002, MM1501, 2N918-3839-3959-4874-5201, SK3039.
A490	BF183, BFX89, BFY78-90, ME3002, 2N918-3572-3959-5130-5180, SK3024.
A492	BF183, BFX89, BFY78-88-90, ME3002, MM1501, SE3001, 2N918-3572-3959-5179, SK3039.
A494	BF167-174-184-194-224-231, GI3392, 2N3932-4134-4876, SK3018.
A495	BF167-174-185-195-224-235, GI3392, 2N3932-4134-4876, SK3018.
A496	BF160-167-196-198-384, MPS6540, 2N2708-3013-3014.
A497	BF159-173-197-199-385, MPS6539, 2N2708-3014-3137.
A610C	E420.
A610S	E420.
A611L	E420.
A611S	E420.
A747B	BC107A-147B-207A-207B, 385A, BCY59B, MPS6575, 2N3242A-3566.
A747C	BC107B-147C-207B-207C-385B, BCY59C, MPS6576, 2N3242B-3566.
A748B	BC108B-148B-172B-208B-386B, GI3710, MPS3710, 2N3710, 40450.
A748C	BC109C-148C-172C-208C-386C, GI3711, MPS3711, 2N3711, 4039 9.
A749B	BC109B-149B-173B-209B-386B, GI3710, MPS3710, 2N3710, 40450.
A749C	BC109C-149C-173C-209C-386C, GI3711, MPS3711, 2N3711, 40399.
A757	BC107A-147A-207A-385A, BCY59A, MPS6575, 2N3242A-3566.
A758A	BC158A-204A-214A, BCY79A, 2N3964-4059-4359, 40419.
A758B	BC158B-204B-214B, BCY79B, 2N3964-4059-4359, 40419.
A759A	BC159A-205A-224A-262A, 2N3964-4059-4359, 40419.
A759B	BC159B-204B-205B-224B-262B, 2N3964-4059-4359, 40419.
A777	BF156-177-257, MM2258, 2N1990, 40372.
A778	BF119-178-258-274, MM2259, 2N3019, 40373.
A779	BF118-179-259, MM2259, SE7055, 2N3742, 40374.
AC105	AC124-131-131/30-153V-173-180-188, GFT31/15, NKT281, OC308, SFT124, TF66/30, 2N43-467-660-1384-2001-2431-4106, SK3004.
AC106	AC124-131-153V-173-180, NKT211, 2N43-243-467-660-1384-2001-4106, SK3004.
AC107	AC125-125R-136-150-151R-160-161-173, NKT229, OC306, 2N413A-1190-1371-1853-2613, SK3004.
AC108	AC122R-126-151V-173-1361, NKT229, OC304I, 2N413A-1191-1371-2429-2613.

AC109	AC122Y-136-151V-173, NKT229, 2N1192-1375-2429-2613.
AC110	AC122G-126-136-151IV-173, NKT225, OC304III, 2N1193-1377-2429-2613.
AC113	AC122-122R-136-151-151IV, NKT223, OC304, SFT353, 2N1191-1352-1371-2613, SK3004.
AC114	AC122-122Y-125-136-151-151IV, NKT229, OC304, SFT353, SK3004, 2N1192-1375-2613.
AC115	AC122R-125-126-136-151-151IV-151V-151VII-160-161-173, BC261B, NKT229, OC304I, 2N1192-1377-2613.
AC116	AC123-126-136-151VI-173-180K, NKT229, SFT353, 2N1189-1190-1192-1371-1377-2429-2613-2907, 2SB75, SK3004.
AC116GN	AC116-125-151V, 2N2907.
AC116Y	AC116-125-151, 2N2429-2907, 2SB75.
AC117	AC117K-128K-153K-180K-188K, NKT304, 2N249-524-659-1384-2001-2431-4106, 2SB370A, SK3004.
AC117R	AC117-128-153.
AC118	AC117-128K-153-180, NKT281, 2N1384-2001-1926-1998.
AC119	AC117-128K-153-180, NKT211, 2N1384-1926-1998-2001.
AC120	AC152-178-180-184, NKT229, SFT323, 2N1008-1131-1384-1999-2431.
AC121	AC121-128K-152-178-180-184, NKT304, SFT323, 2N600-1131-1189-1384-1999-2431,2SB156A.
AC121IV	AC152IV.
AC121V	AC152V.
AC121VI	AC152V.
AC121VII	AC153VII.
AC122	AC136, NKT214, 2N1192-1305-1352-2429-2613, 2SB364, SK3004.
AC122A	AC125-151IV, SFT353.
A122R	AC122-122Y-125-151-151IV, OC304I, SFT353, 2N1190-2613.
AC122Y	AC122-125-151V, OC304, SFT353, 2N1190-2613.
AC122GN	AC122-125-126-151-151VI, SFT353, 2N1190-2613.
AC122V	AC122-126-151VII, OC305I, SFT353, 2N1190-2613.
AC122W	AC122-126-151VII, OC305II, SFT353, 2N1190-2613.
AC122/30	AC126, ASY48, 2SB56A.
AC122/30Y	ASY48V, SFT353.
AC122/30R	ASY48IV, SFT353.
AC122/30R/W	AC122-123, ASY48-77, NKT210, 2N404A-527-586-1190-1191-1371-2163, 2SB460.
AC123	AC136, ACY21, ASY48-48V, 2N527-586-1191-1371-2429-2613, SK3004.
AC123GN	AC123-126, ASY48.
AC123Y	AC123-126, ASY48, 2N2429, 2SB75A.
AC124	AC117-128K-153K-180K, GFT232, NKT302, 2N586-1926-2374-2431-4106, 2SB67.
AC124K	AC128K-153K.

AC125	AC122GN-136-151VI-152-162-170-173, NKT302, OC305II, SFT353, 2N1189-1192-1305-1352-1375-2163-2428-2429-2613, 2SA202, SK3004.
AC126	AC122W-136-151VII-171-173, NKT302, OC305II, SFT353, 2N1189-1192-1305-1375-2429-2613, 2SB383, SK3004.
AC127	AC141-176-179-181-185-186, NKT781, 2N647-1304-1308-2430, 2SD96, SK3010.
AC128	AC117R-153VI-180, NKT302, SFT232, 2N659-1373-1384-1926-2001-2431-4106, 2SB370, SK3004.
AC128K	AC117-117K-128-153K-180K, OC308, 2N4106, 2SB415.
AC129	AC122-125-151, 2SB46.
AC129Y	AC122-125-151, 2SB47.
AC129V	AC151.
AC129BI	AC151.
AC130	AC127-141-186-190, ASY75, NKT734, OC139, SK3010. 2N1302-1605-1808-1993.
AC131	AC117-153VI-180, NKT211, 2N659-1008-1384-1924-2431-3427, 2SB370A.
AC131/30	AC128-153-180, NKT302, SFT232, 2N417-1384-1926-2001.
AC132	AC122-122/30-131-136-162-173-184, NKT213, 2N1008-1008A-1352-1371-1924-2613, 2SB496, SK3004.
AC134	AC121-122-122Y-125-132-151V-151IV-173, NKT224, OC304II, 2N1008-1372, 2613.
AC135	AC121-122-122GN-122Y-125-131-132-151V-151VI-173, NKT219, OC304III, 2N1008-1374-2613.
AC136	AC122Y-125-131/30-132-151V-151VI-162-173, NKT217, 2N1189-1375-2613.
AC137	AC122/30-128-151VII-171-173, NKT229, 2N1192-1375-2613.
AC138	AC122R-124-125-128-131/30-151VI-152-153-173.180 AM54, NKT229, SFT353, TA2063, 2N456-1008-1189-1924-1998-2613.
AC139	AC117R-122Y-124-125-151V-153-153V-180, NKT211, TA2065, 2N1384-2374-3427.
AC141	AC175-176-181-AM72, NKT781, SFT377, 2N1605A-2430.
AC142	AC117-117R-124-153-173-180, AM74, NKT302, 2N586-1926-1998-2001.
AC150	AC122-125R-126-136-151R-173, NKT223, 2N1352-1372-1384-1707-2429, SK3030/3004.
AC150/30	AC125R.
AC150Y	AC107-125-150Y-151V-151VR-161-163, OC306II, SFT353, 2N2613, 2SB73.
AC150GN	AC107-125-151IVR-161-163. OC306III, 2N652-2613
AC151	AC126-136-173, NKT219, 2N652-1193-1375-2429-2613, 2SB439, SK3004.
AC151IV	AC122R-125-126, OC304I, 2N238, 2SB459.
AC151IVR	AC128-150-150Y-151IV-161, OC306I, 2SB73.
AC151V	122Y-125, OC304II.
AC151VR	128-150-150Y-161, OC306II.
AC151VI	AC122GN-125, OC304III.
AC151VIR	AC128-150-150GN-161, OC306III.

84

AC151VII	AC122V-122V/W-125-126, OC305I/II.
AC151VIIR	AC160-160V-161, OC307.
AC152	AC128-131/30-136-180-184, NKT304, 2N600-1375-1924-2431-2613-4234, 2SB156, SK3004.
AC152IV	AC124R-128, OC307-308, SFT322, 2N610-4106, 2S32.
AC152V	AC124R-128, OC307-308, SFT322, 2N610, 2S32.
AC152VI	AC117R-128, OC307-308, SFT322, 2N610, 2S32.
AC153	AC117R-180, NKT211, SFT232, 2N659-1384-2000-2431-3427, SK3004.
AC153K	AC117-117K-128-128K-180K, 2N4106, 2SB415, SK3004.
AC153V	AC124R-128, ACY12-24, ACZ10, 2N467-4106, 2SB222.
AC153VI	AC117R-128, ACY24, ACZ10, ASY12, 2SB415.
AC153VII	AC117R-128, ACY24, ACZ10, ASY12.
AC154	AC117-131-152-152IV-152V-180, NKT211, OC309, SK3004, 2N659-1384-1924-2000-3427.
AC155	AC122-122R-122Y-125-151IV-151V-173, NKT229, OC304I-304T, 2N1190-1191-1192-1371-2613, SK3004.
AC156	AC122-122Y-125-151-151V, NKT223, OC304III, SFT353, 2N1190-1191-1192-1371-2163, SK3004.
AC157	AC175K-176-181-185-186, NKT743, 2N1605A-2430, SK3010.
AC160	AC125R-136-151R-151IV-161-173, NKT225, 2N1191-1352-1375-2613.
AC160A	AC125-151R-151IV/V-161, OC306I/II-307.
AC160B	AC126-151R-151VI/VII-161, OC306III-307.
AC160K	AC125R.
AC160R	AC125-151IVR-161, OC306I, 2SB73.
AC160Y	AC151VR-161, OC30II.
AC160GN	AC151IVR-151VJR-161, OC306III.
AC161	AC125R-136-151V-151VIR-160-173, NKT227, OC306III-307, SFT337, 2N1191-1352-1377-2613, SK3004.
AC162	AC125-128-136-170-173, NKT229, SFT353, SK3004, 2N1191-1192-1377-2429-2613, 2SB459.
AC163	AC126-136-163-171-173, NKT229, 2N1190-1192-1371-2163-2429-2613-2907, 2SB459.
AC165	AC122GN-122-122Y-125-151V-151VI, 173-222, BC261, NKT227, 2N1191-1352-1371-2163, SK3004
AC166	AC117-131/30-152-152V-180-184, NKT211, SK3004, 2N659-1189-1384-1926-2000-4106.
AC167	AC117-131/30-152-152V-180-184, NKT211, 2N659-1089-1189-1384-2000-4106, SK3004.
AC168	AC175K-176-179-181-185-186, 2N647-1012-1473-2430, SK3004.
AC169	AC122-125-151, NKT229, SFT353, 2N1191-1352-1357-2613, SK3004.
AC170	AC122-125-151VI-162-173, NKT281, OC309, SFT353, 2N1191-1192-1373-2613, 2SB54-77A.
AC171	AC122-151VII-163-173, NKT225, SFT353, 2N413A-1190-1194-1377-2163, 2SB460.

AC172	AC163-175-181-183, NKT734, 2N1473-1808 2430, SK3010.
AC173	AC117-128-131-132-152-180, NKT281, 2N1176B-1189-1384-1926-2000.
AC174	AC117-128-131-131/30-153-153V-180, MJ2060, NKT281, 2N117B-1384-1926-2001.
AC175	AC176K-181-187-187K, SK3020, 2N1306-1732-2291-2430-2707.
AC175K	AC176K-183K.
AC176	AC181-186-187, NKT734, 2N1306-1732-2297-2430, SK3010.
AC176K	AC175-176-181-186-187K, 2N2297.
AC177	AC117-131-132-152-152V-173-180, ASY80, NKT211, 2N1176B-1384-2000-3427-4030.
AC178	AC117-122K-128K-153K-180-180K, NKT211, 2N1176B-1384-2001-3427, 2SB370.
AC179	AC175-176K-180-181-187L-187K, NKT713, 2N1605A-2430.
AC179K	AC181K.
AC180	AC117-128-131/30, NKT11, 2N1176B-1185-1384-1926-2000-4106.
AC180K	AC128K-131/30-153, 2N1185-4106.
AC180K/L	AC117-153K-188K.
AC181	AC175-186, NKT734, TLS60, 2N1473-1808-2430.
AC181K	AC176-186-187K.
AC181K/L	AC175-176K-187K.
AC182	ASY26, NKT229, 2N586-651A-1371.
AC183	AC127-176-181-186, NKT734, 2N1308-1732-2430.
AC184	AC128-131-152-153, MJ2060, NKT11, 2N1176B-1384-2000-3427-4106.
AC185	AC127K-176K-183-186, 2N1308-1732-2430.
AC186	AC176-185-187-187K, NKT734, 2N1308-1732-2430.
AC187	AC187K,
AC187K	AC175-175K-176K.
AC188	NK211, 2N1176B-1384-2000-3427, SK3004.
AC188K	AC117-153K, 2N4030, 2SB370.
AC191	AC122-125-125R-150-151R-161, NKT229, 2N1191-1352-1371-2613, SFT353.
AC192	AC122-125-151-151IV-162-170-173, NKT227, SKT353, 2N1191-1352-1375, 2613-2614.
AC193	AC117-117R-128-153-180-184, NKT211, 2N1176B-1384-1926-2000.
AC193K	AC117-128K-153K-180K-188K.
AC194	AC175-176-179-181-186-187-187K, NKT713, 2N1605A-2430.
AC194K	AC175-176K-187K.
ACY16	AC117-128K-180K, ACY33-33V, ASY14, NKT302, SK3009, 2N526-1997-2303-2801.
ACY17	AC117-184K, ASY48, BC157, SK3004, 2CY17. 2N652A.
ACY18	AC117-153V-180K, BC158, SK3004, 2N652-3467, 2SY18.
ACY19	AC117-128-153, 2CY19, 2N249-4106, 2S56, SK3004.

ACY20	AC48V-117-128-131/30-153-184K, 2CY20, 2N1176B-2565-3467, SK3004.
ACY21	AC117-123-180K, ASY48VI, SK3004, 2CY21, 2N1176B-2801-3467.
ACY22	AC117-180K, ACY16-33V, SK3003, 2CY22, 2N651.
ACY23	AC117-122/30-125-180, NKT225, 2N526-1039-1190-1352-2613, SK3004.
ACY24	AC117-128K-184K, ASY14-48V, NKT217, 2N625-1384-1924-2303-2801.
ACY27	AC122-125-126-151-173, ACY23V, NKT225, SK3004. 2N525-651-1191-2447-2613-3467, 2S13.
ACY28	AC122-126-151, ACY23V, NKT225, SFT353, SK3004, 2N526-1192-1377-2613.
ACY29	AC122-125R-150-160-173, ACY32, NKT225, SK3004, 2N237-1191-1352-1371-2613.
ACY30	AC117-122/30-126-128-131/30-153-180, ACY23, 2N237-526-1191-1352-1373-2613, SK3004.
ACY32	AC125R-150-173, ACY29, NKT225, SK3004, 2N527-1191-1352-1377-2613.
ACY33	AC117-180, ACY16K-30-128, ASY14, NKT211, 2N527-618-652-1384-2303-2801.
ACY34	AC117-122-125-128K-180K, ACY16-22-23V, ASY14, NKT229, SK3004, 2N651-1190-1377-2613.
ACY35	AC16-117-122-125-128K-180K, ACY23V, ASY14, NKT225, SK3004.
ACY36	AC117-122-128K-180K, ACY16-23V, ASY14, NKT225, SK3004, 2N1190-1352-1375-2613.
ACY38	AC125R-151-160/B, ASY27, NKT227, 2N1191-1377-2613.
ACY39	ACZ10, 2N652-2303-2801.
ACY40	AC122-173, ACY16-23-29, SK3005, 2N586-625-1189-1371.
ACY41	AC117-131/30-180, ACY23-30, SK3004, 2N586-652-1189-1375.
ACY44	AC117-180, ACZ10, ASY48, NKT217, SK3004, 2N1189-1377-1997.
ACZ10	AC117-128/K-153V-184K, ACY24K, ASY12-48, NKT211, 2SB222, 2N467-2303-2801-4106.
AD103	AD138/50, ADY23, ASZ16, AT1138, MP2008A, NKT402, SFT265-266, SK3014, 2N513A-2869.
AD104	AD134-138/50, ADZ11, ADY23, ASZ16, AT1138A, MP2008A, NKT402, SFT265-266, SK3009, 2N514A-2869.
AD105	AD135, ADY28X, AT1138B, AUY24-28-30, MP2008A, SFT266, SK3009, 2N513B-2870.
AD130	AD129-143-150-153, ADY23, AUY33, MP1529, NKT451, SK3009, TI3029, 2SB337-426, 2N514-1529-2148.
AD130III	AD149/IV, ASZ15, AT1138B, AUY34, OC26-603, MP1552, SFT213-268, TI3029, 2N513B-514B-1100-1146C-1906-2063-2869-3615.
AD130IV	AD149/IV, OC26, SFT213, TI3029, 2N257-2064-2869-3615.

AD130V	AD149/V, OC26, SFT213, TI3029, 2N257-2064-2869-3615.
AD131	AD138/50-142-149, ADY24-28X, ADZ12, ASY16, AUY32-33, MP1530, NKT405, SK3009, 2SB425, 2N514A-1530-2148.
AD131III	AD138/50-149/IV, OC603, SFT250, TI3029, 2SB471, 2N2065-2870-3615.
AD131IV	AD138/50-149IV, SFT250-4012, TI3029, 2N268-2066-2870-3615.
AD131V	AD138/50-149V, SFT250-4012, TI3029, 2N268-2066-2870-3615.
AD132	AD142, ADY23-28Y, ADZ12, ASZ12, AUY28-32, MP1531, NKT403, SK3009, 2SB424, 2N514B-1531-2147.
AD132/2	OC26, OD603, SFT250, 2N2065.
AD132III	ASZ15, AUY22III, OC26-28, OD603/50, SFT250, TI3030, 2SB472, 2N2065-2870-3616.
AD132IV	AUY22IV-28, OC26, SFT250-4012, TI3030, 2N268-2066-2870-3616-4012.
AD132V	AUY22IV.
AD133	AD138/50, ADY23, ADZ72, ASZ16, AT1138, NKT402, SFT266, SK3014, 2N278-513-1549-2869.
AD133III	ADZ11, AUY29III, 2SB236, 2N1549-2869.
AD133IV	AUY29IV, 2N2869-3146.
AD133V	AUY29V, 2N2869.
AD134	AD138/50, ADY23, ADZ11, ASZ11-16, AT1138A, NKT402, SFT265, 2N402-513A-1099-2869.
AD135	ADY24-28X, ADZ12, AUY28-30, AT1138B, NKT403, 2N513B-1099-2870.
AD136	AD160, ASZ18, AT216, NKT402, 2N173-513B-1039-1184B.
AD138	AD143, ADY23, ADZ11, ASZ16-18, MP2060, NKT404, SFT266, SK3014, TI3027, 2N512-1022A-1073A-2869.
AD138/50	AD135-142, ADY28Y, ADZ12, ASZ15-18, AUY22-32, MP2062, NKT403, OC28, SK3014, TI3029, 2N513B-1073B-2870.
AD139	AD143-150-153-62, ADY24-27-28X, NKT451, SK3009, 2N178-513-250A-2869.
AD140	AD150, ADY27, AT1833, AUY34, NKT402, SK3009-3014, 2N513A-1022-1184B-1540-2138-2836-2870.
AD142	AD132-138/50-149, ADY23-28Y, ADZ11, AM92, ASZ15, AUY22-28-32, MP1531, NKT401, OC28, SFT240, TI3029, 2N513B-1073A-2869-2870-3146.
AD143	AD133-138-138/50-150-153, ADY23, ADZ11, ASZ16-17, AM91, AUY33, MP2060, NKT451, SK3014, TI3027, OC28, 40254, 2N512-1073A-2869.
AD145	AD133-133V-138-149-153, ADY23, ASZ16, AUY33, NKT451, TI3027, 40254, 2N307A-513.
AD146	AD166.
AD148	AD139-143-150-163, ADY24, AUY33, NKT452, TI3027, SK3009, 2N514-2142A.
AD148IV	AD139-148-148IV-156, 2SB426, 2N456.
AD148V	AD139-148-156.

AD149	AD140, ADY27-28Y, NKT405, TI3028, SK3009-3013, 40051, 2SB425, 2N514A-1530-2143A-2836.
AD149/01	AD149, SK3009, 2N2836.
AD149/02	AD149, SK3009, 2N2836.
AD149IV	AD149-166, TI156, 2SB471, 2N456-1022-2869-3615.
AD149V	AD149-166, TI156, 2SB471, 2N1022-2836-2869-3615.
AD150	AD143-153, ADY24, AUY33, NKT451, SK3009-3030, TI3027, 40462, 2N514-1142-1530-2142A-2836.
AD150IV/V	AD150-166, TI156, 2SB426, 2N456-2869-3614.
AD152	AD140-162/V, ADY25-28X, MP2060, NKT405, SK3009, TI156, 2SB368, 2N514A-1539-2836-2869.
AD153	AD131-138-140-150-153, ADY27, AUY33, MP2060, NKT402, TI156, 2N513A-1540-2138-2570-2859.
AD155	AD145-162V, ADY25-28X, 2SB367, TI3027, 40254, SK3082, 2N307A-513.
AD156	AD139-140-148-149-152-153, ADY27-28Y, NKT452, SK3009, 40051, TI3028, 2N513A-2142A.
AD157	AD140-149-153, ADY27-28X, NKT405, SK3009, TI3028, 40462, 2N513A-2143A.
AD159	AD136/IV-159, ASZ17, NKT402, SK3009, 2N173-513B-1039-1184B.
AD160	AD136/IV, ASZ16, AT216, NKT402, 2N173-513A/B-1039-1184B.
AD161	AD165, MJ3101, SK3052, 2N1218-1292-1722-4077.
AD162	AD143-152, ADY27-28X/Y, MP2060, NKT406, SK3052, TI3027, 2N1539, 2SB367.
AD163	ADY25, ASZ15, AUY32-34, MP1613, NKT420, TI3030, 2N1362-1542.
AD164	AD162, MP1529, NKT451, 40254, 2SB367, 2N1529-1539.
AD165	AD149-161, MJ3101, 2N1218-1292-1722-4077.
AD166	AD140-149, AU102, BD132, DTG601, MP1531, NKT402, TI3028, 2N512A-2148.
AD167	AD142-149, AU102, DTG601, MP1532, NKT401, TI3029, 2N512B-2147.
AD169	AD143-149-152-162, ADY27, AUY31, NKT405, TI3028, 2SB368, 2N513A-2143A-2148.
AD262	AD139-145-149-162-143-150-153, ADY27, AUY33, NKT405, SK3082, TI3028, 2N513A-2143A-2148.
AD263	AD140-166, ASZ17, AU102, BD132, DTY601, MP1531, NKT402, SK3082, TI3028, 2N513A-1073-2148.
ADY22	AD138-138/50, ADY28Y, ASZ15, AT1138B, AUY22, ADY29, SK3029, TI3029, 2N513B-1099-1146B-2870-3617.
ADY23	AD138-138/50, ADY23-28Y, ADZ12, ASZ15, AUY32, AT1138B, SK3009, TI3029, 2N513B-526-1099-1146B-2870-3617.
ADY24	AD138/50, ADY28X, ASZ15-18, AT1100B, AUY32, SK3009, TI3029, 2N513B-1099-1146B-2870-3617.
ADY25	ASZ18, AT1138B, AUY28-34, NKT420, TI3030, 2N513B-1146C-3618.
ADY26	ADY25-28, AT1138B, AUY28-32-34, SK3012, 2N513B-1146B-1365.

ADY27	AD149, ADY28, AUY31, NKT405, SK3004, TI3028, 2N514A-2143A, 40051.
ADY28	AUY28, SK3009.
ADY28X	AD143, ADY25, AUY28-32-34, ASZ12, NKT451, 2N178-250A-514B-2870.
ADY28Y	AD143, ADY25, ASZ12, AUY28-32-34, NKT451, 2N178-250A-513B-2870.
ADZ11	AD134, ADY25, AT1138B, AUY28-31, MP1550, NKT402, SK3012, SFT240, 2N514A-3615.
ADZ12	AD135, ADY25, AT1138B, AUY28-30, MP1551, NKT403, SFT240, SK3012, 2N514B-1146A/B-3615.
AF101	AF106-125-133-137-144-185-188-190-197, BFX48, BSW19-72, NKT603F, SK3005, TIS37, 2N2273-2495-3127-3324-4916-5354.
AF102	AF121-143-178-190, BFX48, BSW19-72, FT1740, GMO761, NKT613F, SK3006, 2G101, 2N331-499-2273-3324.
AF105	AF126-134-144-185-190-196-197, BFX48, BSW19-73, NKT603F, SK3008, TIS37, 2N2273-2495-3127-3324-4916-5355.
AF105A	AF126-132-137-138-190-196-197, SK3006, 2N2495.
AF106	AF143-178-186-190, AM13, BFX48, BSW19-72, GM378A, GMO761, NKT613F, SK3006, 2G402S, 2N2273-2496-2635-3324-5354.
AF107	AF109R-121S-142-190, BSX48, BSW19-72, GMO760, NKT613F, AFY11, 2N1141-1195-3281-3324-4916-5354.
AF108	AF109R-121S-142-190, AFY11, BFX48, BSW19-72, GMO760, NKT613F, SK3006, 2N1195-3280-3324-4917-5354.
AF109	AF109R-139-190, BFX48, BSW72, GMO760, FT1746, MEO491, NKT674F, SK3006, 2N2273-3324-3883-5354.
AF109R	AF139-180-190, BFX48, BSW19-72, NKT674F, 2N2273-3127-3324-3883-5354.
AF110	AF190.
AF111	AF135-172-185-197, BFX48, BSW19-73, FT1746, NKT603F, 2N499-2273-3324-4034-5354.
AF112	AF144-196, BFX48, BSW19-73, FT1746, NKT603F, 2N1110-2273-3324-4034-5354.
AF113	AF135-136-143-196, BFX48, BSW19-73, NKT613F, 2N2273-2495-2635-3324-5354.
AF114	AF144-194, BFX48, BSW19-73, GM1213B, NKT613F, SK3006, TIS37, 2G402S, 2N2273-2495-3127-3324-4916-5354.
AF115	AF146-185-195, BFX48, BSW19-72, GM1213B, NKT603F, SK3006, TIS37, 2G401S, 2N2273-3127-3324-4916-5354.
AF115N	AF125-135-146, BFX48, BSW19-72, NKT603F, TIS37, 2N2273-3127-3324-4916-5354.
AF116	AF135-136-146-185-196, BFX48, BSW19-73, GM1213B, NKT613F, SK3006, TIS37, 2G401S, 2N2273-3127-3324-4916-5354.

AF116N	AF126-134-146, BFX48, BSW19-73, NKT613F, TIS37, 2N2273-3127-3324-5355.
AF117	AF136-146-197, BFX48, BSW19-73, GM1213B, NKT603F, SFT354, SK3006, 2N2273-2635-3324-4916-5354.
AF117N	AF127-137-146, BFX48, BSW19-73, NKT603F, 2N2273-2635-3324-4916-5354.
AF118	BFW20, SFT162, SK3006, 2N2207-3251A-3798-3963.
AF119	AF185.
AF120	AF185, SK3006.
AF121	AF102-139-178-201, BFX48, BSW19-73, FT1746, GM378A, NKT613F, SK3006, 2N499-2273-2495-3324-3588-5354.
AF121S	AF143-202S, BFX48, BSW19-73, FT1746, 2N499-2273-3324-5354.
AF122	AF178, GMO761, 2N2495.
AF124	AF144-178-194, BFX48, BSW19-72, GM1213B, NKT674F, SK3006, 2SA235, 2N1524-2273-2495-2635-4035-4916-5354.
AF125	AF131-135-136-142-185-195, BFX48, BSW19-73, GM1213B, NKT674F, SK3006, 2N991-1110-1525-2273-2635-2411-3324-4916-.
AF126	AF136-143-196, BFX48, BSW19-73, GM1213B, NKT674F, SET316, SK3006, 2N499-1526-2189-2411-3383-4916-5355.
AF127	AF136-146-185-197, BFX48, BSW19-73, GM1213B, NKT613F, SK3005, 2N993-1527-1748-2191-2412-3325-4034-4916-5354.
AF128	SK3005, 2N1303.
AF129	AF149, BFX48, BSW19-73, NKT613F, SK3006, 2N1141-1524-2273-2279-3127-4916-5354.
AF130	AF144-194, BFX48, BSW19-73, NKT613F, SK3006, 2N346-1525-2495-2635-3127-3323-4035-4916-5354.
AF131	AF134-146-185-195, BFX48, BSW19-73, NKT677F, SK3006, 2N1525-2189-2412-3323-3324-4916-5354.
AF132	AF136-138-146-196, BFX48, BSW19-73, NKT674F, SK3006, 2SA155, 2N641-1526-2190-2412-3323-3324-4916-5354.
AF133	AF136-147-197, BFX48, BSW19-73, NKT674F, SK3006, 2N502-641-1527-2191-2412-3325-4034-4916-5355.
AF134	AF129-144-197, BFX48, BSW19-73, NKT613F, SK3008, 2N1524-2273-2635-2495-3323-4035-5355.
AF135	AF146-195, BFX48, BSW19-73, SK3008, 2N1525-2188-2412-3279-3324-4917-5355.
AF136	AF125-132-146-196, BFX48, BSW19-72, NKT677F, SK3008, 2N1526-2189-2412-3284-3323-4034-4916-5354.
AF137	AF120-147-197, BFX48, BSW19-72, NKT613F, SFT316, SK3008, 2N1527-2190-2273-3284-3323-4916-5354.
AF138	AF133-146-179-197, BFX48, BSW19-73, FT1746, SK3006, 2N1526-2190-2273-3284-3323-4034-5355.
AF139	AF186, AM18, BFX48, GM290A, GMO290, M9031, MM139, 2N2244-3279-3323-3324-5354.

AF142	AF112-130-134-194, BFX48, BSW19-72, NKT674F, 2N1524-2188-2273-2411-3323-4035-4916-5354.
AF143	AF125-131-135-195, BFX48, BSW19-72, NKT613F, 2N1525-2190-2273-3324-4916-5354.
AF144	AF125-126-132-134-136-137-196, BFX48, BSW19-72, NKT603F, SK3006, 2N1525-2191-2273 4916.5354.
AF146	AF125-131-135-195, BFX48, BSW19-72, NKT613F, 2N1526-2191-2273-3324-4916-5354.
AF147	AF106-126-127-133-136-137, BFX48, BSW19-72, NKT613F, 2N1525-2191-2411-5354.
AF148	AF106-126-127-133-136-137-197, BFX48, BSW19-72, FT1746, NKT613F, 2N499-1526-2191-2412-3323-4034-5354.
AF149	AF126-131-132-135-136-196, BFX48, BSW19-73, NKT667F, 2N1838-2273-2873-3323-4916-5355.
AF150	AF127-133-134-137-197, BFX48, BSW19-72, NKT613F, 2N2273-2635-2873-3325-4034-4916-5354.
AF164	AF126-130-132-134-194, BFX48, BSW19-72, NKT603F, 2N1526-2191-2273-3323-4035-5354.
AF165	AF125-131-134-135-136-195, BFX48, BSW19-72, NKT603F, 2N1524-2188-2273-3324-4916-5354.
AF166	AF126-132-136-196, AM11, SFT316, SK3006, 2N1180-3323-4034.
AF167	AF124-134, BFX48, BSW19-72, NKT613F, 2N1524-2200-2635-4916-5354.
AF168	AF125-126-131-135-136-195, BFX48, BSW19-72, NKT613F, 2N1183-1525-2273-2635-3324-4916-5354.
AF169	AF106-126-127-129-132-136-137-196, BFX48, BSW19-72, FT1746, NKT613F, 2N1527-2190-2273-3325-4034-5354.
AF170	AF127-132-136-137-196, AM14, BFX48, BSW19-72, FT1746, NKT613F, 2N1527-1635-1639-2188-2273-3325-4034-5354.
AF171	AF133-135-136-137-196, BFX48, BSW19-73, NKT613F, 2N2188-2273-3324-4035-4917-5355.
AF172	AF127-133-136-137-197, AM12, BFX48, BSW19-72, NKT603F, 2SA432, 2N1638-2273-2635-3324-3325-4034-5355.
AF178	BFX48, BSW72, FT1746, GMO761, NKT674F, SK3006, 2SA432, 2N711B-2273-2492-2495-3324.
AF179	AFY11, BFX48, BSW72, NKT613F, SK3006, 2N711B-2273-2654-3324-4916.
AF180	AF109R-133-200, BFX48, BSW19-72, NKT613F, 2N2273-3074-3127-3325-4913-5354.
AF181	AF121-139, BFX48, NKT674F, SK3006, TI364, 2N1748-2411-2873-3075-4916-5355.
AF182	AF121-201, BFX48, BSW19-72, FT1746, SFT163, SK3006, 2N499-2273-2843-5354.
AF183	AF121-178-182-201, BFX48, BSW72, NKT613F, 2N2273-3324-4916.
AF184	AF109R-178-182-201, BFX48, BSW72, NKT613F, 2N2273-2873-4916.

AF185	AF121S-187, BFX48, BSW72, GMO761, NKT603F, SK3006, 2N2273-2492-2873-4916.
AF186	BFX48, GMO290, GM290A, NKT677F, 2N2188-2200-22,44-2635-2873-3279-4916-5043.
AF186G/W	AF121-189, AFY15, BFX48, NKT613F, 2N1195-2200-2873-3371-4916-5354.
AF187	AF121-133-201, AFY15, BFX48, BSW72, FT1746; NKT674F, SFT307, 2N2244-2273-2635-3324-5044.
AF188	AF121-133, AFY15, BFX48, BSW72, NKT603F, SFT308, 2N2244-2273-3324-4916-5044.
AF189	AF121-124-133-134-139, AFY15, BFX48, BSW19-73, FT1746, NKT613F, 2N1195-2189-2273-3324-5354.
AF190	AF121-124-126-133-136-188, AFY15, BFX48, BSW19-73, NKT613F, 2N2189-2243-3324-4916-5354.
AF191	AF127-137-197, BFX48, BSW73, NKT677F, SFT308, 2N1527-2412-2635-4916.
AF192	AC130, BSY62, BSW82, NKT713, SK3011, 2N1302-1605-1808.
AF193	AF121-126-127-132-136-137 BFX48, BSW72, FT1746, NKT613F, SFT317, 2N1527-2273-3323-4034.
AF194	AF124-125-130-134-135, BFX48, BSW73, NKT674F, SFT358, 2N2273-2635-1524-3323-4035-4916.
AF195	AF125-131-133-135, BFX48, BSW73, NKT674F, SFT357, 2N1525-2273-3323-4916.
AF196	AF125-126-132-135-136, BFX48, BSW73, NKT677F, SFT354, 2N1526-2273-3323-4034-4916.
AF197	AF127-133-137, SFT316B, 2N3325-4034.
AF198	AF127-133-136-138, BFX48, BSW19-72, NKT674F, SFT316VI, 2N1526-2273-3325-4916-5354.
AF200	AF121, BFX48, BSW19-72, NKT674F, 2N2411-3127-3324-3883-4916-5354.
AF201	BSW19-72, FT1746, NKT677F, 2N705-2412-3127-3324-3883-5354.
AF202	AF121S-202S, BSW19-72, FT1746, NKT674F, 2N705-2412-3127-3324-3883-5354.
AF202S	AF121S, BFX48, BSW19-72, NKT677F, 2SA229-240, 2N2273-3324-4916-5354.
AF239	BF272, GM378A, 2N3304-3323.
AF240	AF139-239/S-267-269-280, BF272, 2N700-3323.
AF250	AF239S-267-280;
AF251	AF239/S, BF272 2N700-3279-3324.
AF252	AF239/S-240, BF272, 2N700A-3280-3785.
AF253	AF109R-139-239-267, BF272, 2N502A-3304-3784.
AF256	AF106, BF272, GMO761, NKT674F, 2N2273-3324-4916.
AF257	AF239S-267-279, BF272, 2N3230-3304-3324.
AF264	AF106, BFX48, BSW19-72, NKT674F, 2N3304-3324-4916-5354.
AF267	AF239S-240-279, BF272, 2N3280-3304-3324-5043.
AF268	AF239S, BF272, NKT613F, 2N3281-3304-3323-3324.
AF269	AF239/S-240-280, BF272, NKT603F, 2N3282-3304-3323/4.
AF279	AF179-239S-267-269, BF272, 2N3280-3304-3324-5044.

AF280	AF239/S-240-257-267-269, BF272, 2N3281-3304-3324-5043.
AFY10	AFY16, BFX48, BSW19, FT1746, NKT613F, SK3006, 2N499-2273-3324-5354.
AFY11	AFY16, BFX48, BSW19, NKT603F, SK3006, 2N1141-3304-3324-4916-5354.
AFY12	AF106-121, AFZ12, BFX48, BSW19, NKT603F, 2SA431, 2N3304-3371-3883.
AFY13	AF124-134, BFX48, BSW19-72, NKT613F, 2N1524-2099-2635-4916-5354.
AFY14	AF106, ASY27, BSW72, NKT613F, SFT316K, 2N2415-2873-3702.
AFY15	AF125-126-133-136, BSW19, NKT603F, SFT317, TI397, 2N582-3638-5227-5354.
AFY16	AF139-239S, BF272, M9031, MM1139, 2N711A-1142-1495-4035.
AFY18	AF139-239S-240, AFY16-19, BF272, ME0491, SK3006, 2N711A-1142-1495-3307-4035.
AFY19	AFY16-18, BF272, ME0491, 2N711A-1142-1561-3883-4035.
AFY25	AF126-239S, AFY16-34-42, BF272, MM1139, 2N1141-3283-3324.
AFY26	AF239S, AFY18-34-40, BF272, MM1139, 2N1142-3127-3324.
AFY29	AF126-133-293S, AFY16-18-40, BF272, MM1139, SFT317, 2N1142-3127-3324-5244.
AFY34	BF272, 2N3307-3784-4952-5043.
AFY37	AF239S, AFY18-40, BF272, ME0491, 2N1141-2929-3784.
AFY39	AF239S, AFY18-40, BF272, ME0491, 2N1141-2929-3307-3784.
AFY40/R	AF239S, AFY18-39, BF272, ME0491, 2N1141-2883-2929-3279-3784.
AFY41	AF239/S, AFY18-42, BF272, MM1139, 2N1142-3324-5244.
AFY42	AF239S, AFY18-40-41, BF272, MM1139, 2N1142-3324-5244.
AFZ10	AF239, AFY14K-18-41, BFX48, MM380, NKT613F, TIXM17, 2N3784-4916.
AFZ11	AF106, AFY12, BFX48, BSW19-72, GM378A, NKT613F, 2N2273-3127-3324-4916-5354.
AFZ12	AF106, AFY12, BFX48, BSW19-72, GM378A, NKT613F, 2N1305-2273-3324-4916-5354.
AL100	AU103-105, B1085, DTY600, MP1612, 40440, 2N1044-1073A-1176B-5007.
AL101	AU103-105, DTY601, MP1612A, B1085, 2N1045-1073B-1612A-5009.
AL102	AM114, AD167, B1085, AU103-105, DTY110, MP1612B, 2N1045-1073B-1176C-1906-5003.
AL103	AD166, AM111, AU103-105, B1181, DTY600, MP1613, 2N1045-1073B-1176C-1906-5005.
ALZ10	AD140-149-153, ADY27-28X, NKT405, SK3005, 40394, 2N601-1039.

AM11	SFT316.
AM12	2N1524-1638.
AM13	AF106-184, SK3016, 2N1177.
AM14	2N1526-1639.
AM15	2N1178/9.
AM18	MM139.
AM51	AC191, SK3004, 2N2613.
AM52	2N215-406.
AM53	AC126-161-192, SK3004, 2N2614.
AM54	AC151, SK3004, TA2063.
AM71	40253.
AM72	SFT377.
AM73	2N591.
AM94	40254.
AM251	BC107.
AM252	BC108.
AM253	BC109.
AM254	BC267.
AM255	BC268.
AM256	BC269.
AM257	BC270.
AM258	BC271.
AM259	BC272.
AM260	BC300.
AM261	BCY58.
AM262	BCY59.
AM263/4	BC301.
AM276	BC297.
AM291	2N3055.
AM293	BD142.
AM294	BD141.
ASY12	AC124K, 2N1924.
ASY12-I/II	AC124-128-180, ACY24, ASY70, BSW72, BSX36, NKT229, SFT323, 2N1189-1384-1998-2927.
ASY13	ASY48V, 2N1926.
ASY13-I/II	AC124-128-153, ACY24, ASY81, BFX74A, BSW74, ME0401, NKT217, 2N398B-1924-5022.
ASY14	ASY48IV.
ASY14-I/II/III	ACY24, ASY48-77-81, BCW86, BSW74, NKT217, 2N398/A-1924/5-2890-5022.
ASY24	ASY24-48-56-71-80-81, BCW86, BSW72, NKT217, SK3008, 2N398A-1924-5022.
ASY24B	ASY48-71-76-81, BCW86, BSW72, NKT217, 2N398B-1925-5022.
ASY26	BSW72, BSX36, NKT210, SK3005, 2N653-1192-1307-1371-2613-2927.
ASY27	ASY26-58, BSW72, BSX36, NKT135, SK3005, 2N1192-1304-1371-1384-1707-2927.
ASY28	AC127, BSW82, BSX12, NKT717, 2N1605A-1808-2430-5135.
ASY29	AC127, BSW82, BSX12, NKT713, SK3011, 2N1306-1605A-1808-2430-5135.
ASY30	ASY32-48-61-77-80-81, BFY74A, BSW74, NKT217, 3006, 2N398-1303-1176B-1925-5022.

ASY31	ACY15, AF125, ASY32-56, BSW72, BSX36, NKT210, 2N653-1192-1303-1371-2613-2927.
ASY32	AFY15-32, ASY27-48-56, BSW72, BSX36, NKT210, 2N1193-1372-2613-2927.
ASY48	ASY71-81, BFX74A, BSW74, NKT217, 2N398-1924-2906/A-5022.
ASY50	AC122-131/30, ASY70/V-80-81, BSX36, NKT229, SK3004, 2N526-1189-1307-1384-1998-2927.
ASY52	ACY17, ASY48/V-71, ME0401, NKT217, BFY74A, 2N398B-1307-1925/6.
ASY53	AC127-128-131/30, ASY28-70IV, SK3011, 2N526-1302-1605A.1808.2430-5135.
ASY55	BSX36, NKT135, TF49, 2N1193-1303-1373-1384-1707-2927.
ASY56	ASY27, BSX36, TF49, 2N1193-1303-1375-1384-2927.
ASY57	SK3006, TF49, 2N1303-1305.
ASY58/9	BSX36, NKT135, SK3006, TF49, 2N1303-1305-1373-2613-2927.
ASY60	ASY27, BSX36, NKT135, 2N1192-1373-2613-2927.
ASY61	AC127, BSX12, NKT717, 2N1302-1304-1605A-1808-2430-5135.
ASY62	AC127, BSX12, NKT717, SK3011, 2N1304-1605A-1808-2430-5135.
ASY63	AC131/30, ASY70-80-81, BSX36, NKT229, SFT232, 2N1189-1998-1384-2927.
ASY64	ASY26, BSX36, NKT135, 2N1192-1307-1373-2613-2927.
ASY66	ASY27, BSX36, NKT135, 2N1193-1375-1996-2613.
ASY67	2N2191.
ASY70	AC124K, ACY24, ASY63-81, BSX36, NKT229, OC74, SFT232, 2N404-1926-1998-1384-2927.
ASY71	ACY24, NKT229, 2N398-1998-2042-2890-4928.
ASY72	SK3011, 2N1304.
ASY73	AC127, ASY28, BSW82, BSX12, NKT717, SK3011, 2N1605A-1808-2430-5135.
ASY74	AC127, ASY29, BSW82, BSX12, NKT717, SK3011, 2N1605A-1808-2430-5135.
ASY75	AC127-153V, ASY29, BSW41, BSX12, SK3011, 2SC91, 2N1306-1605A-1808-2430.
ASY76	ACY16/K, ASY48, BCW86, BFS95, BSW72, ME0401, SK3005, 2N398-526-1925-5022.
ASY77	ACY24/K, ASY48-81, BCW86, BFS92, BSW74, ME0401, NKT217, SK3005, 2N398A-526-1926.
ASY78	ACY16, ASY48, BCW86, BSW74, ME0401, NKT217, 2N398B-1307-1925-1998-5022.
ASY80	ACY24/K, ASY48, BCW86, BSW74, ME0401, NKT217, SK3005, 2N398A-527-1307-1925.
ASY81	ACY24/K-77, ASY48, BCW86, BSW74, ME0401, NKT217, SFT232, 2N398B-1926-1998-5022.
ASZ10	ASY19-30K-48-77, BCW86, BSW74, ME0401, NKT317, SFT232, SK3006, 2N398-1925-1926-1998-5022.
ASZ11	ASY26, SK3006.
ASZ12	ASY27.

ASZ15	ADY25, AT1138B, AUY22III-28-34, SK3009, TI3029/30, 2SB341, 2N513B-1146C-3613.
ASZ16	AD138/50, ADY23, AUY21IV-28, AT1138B, SK3009, TI3028, 2SB339, 2N513A-1146-3615.
ASZ17	ADY24, AT1138B, AUY21III-28, SK3009, TI3028, 2SB339, 2N513A-1146B-3615.
ASZ18	AT1138B, AUY21II-28-34, SK3009, TI3029, 2SB340, 2N513B-1146B-3613.
ASZ20	ACY24, ASY48-80-81, BCW86, BSW74, GM1213B, NKT613F, SFT232, 2N308-1920-1998-5022.
ASZ20N	ACY24, ASY48-80-81, BCW86, BSW72, NKT613F, SFT232, 2N2273-3324-5022.
ASZ21	ASY26, BC178VI, BFX48, BSW72, NKT613F, 2N711A-2273-3324-4916.
ASZ23	ASY24, BCY70.
ASZ30	ASY26, SK3006, 2N1925.
AT207	AD136, NKT405, 2N514B-1039-1184B.
AT209	AC173, ACY16-21, ASY48, BSW74, BSX36, NKT229, 2N1189-1384-1998-2927.
AT210	AC173-180, ACY16-23, ASY70, BSW74, BSX36, NKT229, 2N1189-1998-2613-2927.
AT216	AU104-112, MP3731, 40349.
AT318	BF115-167-196-224-334, BFY64, ME1001, MM1803, SE1001, SK3039, 2N2857-5144.
AT319	BF115-167-196-334, BFY64, ME1001, MM1803, NKT10339, SE1001, SK3039, 2N2857-5144.
AT321	BF115-167-197-224-335, BFY65, ME1002, MM1803, NKT10439, SE1002, SK3039, 2N2857-5144.
AT322	BF173-195-235-273A, GI3709, ME2001, MPS3709, NKT12329, SE2001, SK3039, 2N2708.
AT323	BF173-195-235-273B-385, GI3709, ME2001, MPS3709, NKT12329, SE2001, SK3039, 2N2708.
AT324	BF173-194-234-274A-384, GI3710, ME2002, MPS3710, NKT12429, SE2002, SK3039, 2N5189.
AT325	BF167-196-273, GI3709, ME3001, MPS6507, NKT16229, SE3001, SK3009, 2N4874-5189.
AT326	BC148A-208A-386A, BCY58A, MPS3710, GI3710, NKT10419, 40450, SK3039, 2N3710.
AT327	BC148B-208B-386B, BCY58B, GI3710, MPS3710, NKT10519, 40450, SK3039, 2N3710.
AT328	BC148C-208C-386B, BCY58C, GI3711, MPS3711, NKT10519, 40399, SK3039, 2N3711.
AT329	BF167-195-235-255-273-385, GI3709, MPS3709, NKT16229, SE5025, SK3122, 2N2708.
AT330	BF167-194-234-254-274-386, GI3709, MPS3709, NKT16229, SE5025, SK3039, 2N2708.
AT331	BC157-158A-205A-212-291, SK3114, 40406, 2N3245-3906 4121.
AT332	BC158/B-205B-213-260B-292, SK3114, 40419, 2N3244-3964-4359.
AT333	BC158C-159-205C-224-260C-292, 40419, 2N3964-4059-4359.
AT335	BF167-196-273, GI3708, ME3001, MPS6507, NKT16229, SE3001, SK3122, 2N4874-5189.

AT337	BC149C-208C-386C, BCY58C, GI3711, MPS3711, NKT10519, 40450, SK3122, 2N3711.
AT341	BC147A-207A-385A, BCY58A, GI3710, MPS3710, NKT10519-40399, SK3117, 2N3710.
AT342	BF167-195-196-273, GI3708, ME3001, MPS6507, SE3001, SK3117, 2N4874-5189.
AT343	BF167-195-196-273, ME3001, GI3708, MPS6507, NKT16229, SE3001, SK3117, 2N4874-5189.
AT450	AU104-106, MP3731, 40349.
AT605	BD160.
AT650	AD149-150, ADY27-28Y, AUY31, NKT405, 2N513B-1365-2870.
AT1138	ADY23, ADZ11, ASZ16, AUY21-28-31, NKT402, 2N512-2869-3615.
AT1138A	ADY24, ADZ11, ASZ16, AUY21-28-31, NKT402, TI3028, 2N512B-2869-3615.
AT1138B	ADY25, ADZ12, ASZ15, AUY22-28-32, NKT401, TI3029, 2N513B-2870-3615.
AT1833	ADY24, ADZ11, ASZ17, AUY21-28-31, NKT402, TI8027, 2N513-2869-3615.
AU101	AUY22, AU105, B1085, DTG2300, MP1612A, SK3014, 2N1100-1543-1906-5003.
AU102	AUY21III, DTG110A, SK3009, TI3027, 2N1184B-1359-3312.
AU103	AU104, B1085, DTG110A, MP1612B, TI3031, 2N1364-1906-5007.
AU104	AU103, B1178, BD299, DTG1110, MP1612B, TI3031, 40440, 2N1073B-5009.
AU105	AU103, ASZ15, B1178, DTG1110, MP1612A, TI3031, 2N1364-1906-5005.
AU106	AU111, BD299, DTG1010, MP3730, 40439, 2N5325.
AU107	AU103-104-111, DTG1110, MP3731, 40440, 2N5324.
AU108	AU101-105, ASZ15, DTG2300, MP1612A, 2N908-1073B-1906-5005.
AU110	AU101-103, DTG1110, B1178, MP1612B, TI3031, 2N1364-1906-3442-5007.
AU111	AU106, DTG1010, MP3730-40439, 2N5325.
AU112	AU106, DTG1010, MP3730, 2N5325.
AUY10	AD138/50, ADY22-28, ADZ12, AT1138B, AUY22, NKT401, SK3009, 2N513B-1099-2870-4033.
AUY18	AD138/50-148, ADY23, ADZ11, ASZ15, AUY18-31, AT1138A, BD132, NKT404, TI3029, 2N3615-5134.
AUY19	AD138/50, ADY24, ADZ11, ASZ15, AT1138A, AUY31-33, NKT403, TI3028, SK3014, 2SB338, 2N157A-513A-2869-3613.
AUY20	AD132, ADY24, ASZ15, AUY28-32, ADZ12, AT1138B, NKT401, SK3014, TI3029, 2SB341, 2N157A-513B-2870.
AUY21	ADY23-28, ADZ11, ASZ15-16-18, AT1138A, AUY28-34, MP2061, NKT402, SK3009, TI3028, 2SB339, 2N513A-2869.
AUY21A	ASZ15.
AUY22	ADY23, ADZ12, ASZ16, AT1138B, AUY20IV-28-32, MP2062, NKT401, SK3009, TI3029, 2N513B-2870.

AUY22A	ASZ15.
AUY28	ADY25-26, ASZ15, AUY22III-30-32-34, AT1138B, MP2063, NKT401, TI3030, 2SB340, 2N513B-2870.
AUY29	ADY23, ADZ11, ASZ16, AT1138A, AUY21-29IV-31, MP2061, NKT406, TI3028, 2N513A-2869.
AUY30	ADY22, ADZ12, ASZ15, AT1138B, AUY22/II-28, MP2062, NKT403, TI3029, 2N513B-2870.
AUY31	ADY23, ADZ11, ASZ16, AUY21/III-28, NKT404, MP2061, TI3028, 2N513A-2869.
AUY32	ADY24, ADZ12, ASZ15, AT1138B, AUY20-22-28, MP2062, TI3029, 2N513B-2870.
AUY33	ADY24, ADZ11, ASZ16, AT1138A, AUY19III-21-28, MP2061, NKT402, TI3028, 2N513A-2869.
AUY34	ADY24, ADZ12, ASZ15, AT1138B, AUY28-30, MP2063, NKT420, TI3030, 2N513B-2870.
AUY35	ASZ16, AUY30.
AUY37	ADY25, ASZ15, AUY30-34/II, AT1138B, MP2063, NKT420, TI3030, 2N513B-2870.
AUY38	ASZ15.
AUZ11	AD140-149-150-153, ADY27-28X, ASZ16-17, AUZ10, NKT451, 40462, TI3028, 2N514A-2143A.
B1085	AU101-103-105, DTG1110, MP1612A, SK3014, 2N1543-1100-1906-5005.
B1178	AU103-105-110, DTG2400, MP1612B, TI3031, 2N1364-1906-5007.
B1181	AU103-105-110, DTG1110, MP1612B, TI3031, 2N1364-1906-5009.
B10474	ADY24, ADZ11, AT1138A, AUY21-28-31, MP2061, SK3014, TI3028, 2N513A-2869.
B10475	AU101-104-110, DTG1110, MP1612A, SFT240, SK3014, TI3031, 2N1364-1906-5005.
B10912	ADZ11, AT1138A, AUY21-28, DTG110B, MP2061, NKT406, TI3028, SK3009, SFT266, 2N513A-2869.
D10913	ADZ12, AT1138B, AUY22-28, DTG110B, MP2062, NKT401, SFT267, SK3009, TI3029, 2N513B-2870.
B102000	ADZ11, AT1138A, AUY21-28, DTG600, MP2061, NKT402, SFT265, TI3028, 2N513A-2869.
B102001	ADZ11, AT1138A, AUY21-28, DTG600, MP2061, NKT402, SFT265, TI3028, 2N513A-2869.
B102002	ADZ12, AT1138A, AUY21-28, DTG600, MP2062, NKT401, SFT266, TI3029, 2N513B-2870.
B102003	ADZ12, AT1138B, AUY22-28, DTG601, MP2062, NKT401, SFT267, TI3029, 2N513B-2870.
B103000	ASZ16, AT1138A, AUY29, DTG600, MP2061, NKT402, SFT265, TI3027, 2N513A-2869.
B103001	ADZ11, AT1138A, AUY21-28, DTG600, MP2061, NKT402, SFT265, TI3028, 2N513A-2869.
B103002	ADZ12, AT1138B, AUY22-28, DTG601, MP2062, NKT401, SFT266, TI3029, 2N513B-2870.
B103003	ADZ12, AT1138B, AUY22-28, DTG602, MP2062, NKT401, SFT267, TI3029, 2N513B-2870.
B103004	ADZ12, AT1138B, AUY22-28, DTG602, MP2062, NKT403, SFT268, TI3029, 2N513B-2870.

B113000	AU101-105, DTG2300, MP1612A, 2N908-1073B-1906-5005.
B113001	AU103-105-110, DTG2400, MP1612B, TI3031, 2N1364-1906-5009.
B113002	AU103-104-110, DTG1110, MP1612B, TI3031, 2N1364-1906-5009.
B113003	○ AU103-104-105, DTG2400, MP1612B, TI3031, 2N1364. 1905-5007.
B113004	AU103-105-110, DTG1110, MP1612B, TI3031, 2N1364-1906-5009.
B113005	AU103-104, DTG1110, MP1612B, TI3031, 2N1364-1906-5009.
BC100	BCY49A, BD115-127.
BC107	AM251, BC147-167-207-317, MPS6566, SK3020-3122, ZTX107; TT107.
BC107A	BC171A-182A-207A-237A-385A, MPS6566-6575, SK3020-3122, 2SC458, 2N324A-2921-3566-3568.
BC107B	BC147B-167B-171B-182B-207B-237B-385B, MPS6566-6576, SK3122, 2N3242B-3566-3568.
BC108	AM252, BC148-168-208-318, MPS6520, SK3020, ZTX108, TT108.
BC108A	BC148A-168A-172A-183A-208A-238A-368A, GI3710, BSY76, MPS6530-3710, SK3122, 40450, 2SC458, 2N3391-4135.
BC108B	BC148B-168B-172B-183B-208B-238A-386B, BSY80, GI3710, MPS3710-6520, SK3020-40450, 2N4134.
BC108C	BC148C-168C-172C-183C-208C-238C-386C, GI3710, MPS3711-6520, SK3122, 40399, 2N3711.
BC109	AM253, BC149-169-209-319, MPS6521, SK3020, ZTX109, TT109.
BC109B	BC149B-169B-173B-184B-209C-239B, GI3711, MPS6521-6571, SK3020. 40397, 2SC458, 2N5126.
BC109C	BC149C-169C-173C-184C-209C-239C, GI3711, MPS6521-6571, SK3122, 40450, 2N5132.
BC110	BC117-174, BF177, BSW65, BSY55, ME8002, SK3122, 2SC856, 2N699-1990-3701-4410-4961-5184.
BC112	BC146-246, SK3039.
BC113	BC148B-168B-172/B-183/B-208B-238/B-386B, GI3707, MPS3707-6520, NKT10519, SE4002, SK3122, SX3707, 40450, 2SC458, 2N3568.
BC114	BC149B-169B-171-173/B-184/B-207A-209B-237B-238B-239/B, GI3711, SK3122, 40397, MPS3711-6521, ZTX114, 2N3568.
BC115	BC182-237A, BCY58A, GI3706, MPS6566, NKT10439, SE6001, SK3122, 2SC984, 2N3568-3704-3933-4123-4876.
BC116	· BC117-160/6-177-212/LA-261-307VI-313, GI3644, MPS6534, NKT20339, 2SA565, 2N2904-2906-3502-3468-4037-4403-5041.
BC116A	BC177-212/LB-261, 2N2904-5041.
BC117	BC141-145-178; BF117-178-257-297, BFR86, MM2258, SE7002, 40373, 2N3019-5831.

BC118	BC147-167-129-171-182-207/A-237/A-385A, MPS6566-6575, NKT10439, SK3122, 40450, 2N3566-3568-3694-4401.
BC119	BC140-211, BSX46VI, SBC119, SK3024, 2N1507-2218-2410-3252-5189.
BC120	BC108-140-211-286, BFY50, BSX45VI, SK3024, MM3725, 2N1507-2102-2218-2410-3036-3252/3-3665-5189.
BC121	BFY87A, SK3020.
BC121W	BC146/R, DFY07Y, 2SC458.
BC121BL	BC146 Blue.
BC121GR	BC146 Green.
BC121YEL	BC146 Yellow, BC107A.
BC122	BC146, BFY87A, SK3020.
BC122W	BFY87A, BC146/R, 2SC458.
BC122Y	BC107A-146Y.
BC122B	BC146G.
BC122G	BC146G.
BC123	BC112, SK3020.
BC123W	2SC458.
BC123Y	BC107A.
BC125	BC107A-171-182-207A-237A-337-385A, GI3705, MPS6566-6575, SK3122, 2N3242A-3566-3705.
BC126	BC160VI-177-212-261-307-313-338-360VI, GI3644, SK3114, TIS93, MPS6535, 2N3638A-3703-4037-4402-5041-5138.
BC127	BC146R, SK3039.
BC128	BC146G, SK3039.
BC129	BC107-147-167-171-182-207-237-317-385, MPS6566-6575, NKT10519, SK3122.
BC129A	BC107A-182A, MPS6566, 2SC281, 2N3568.
BC129B	BC107B-182B, MPS6566.
BC130	BC108-148-168-172-183-208-238-386, GI3711, MPS3711-6520, NKT10419, SK3122, 40399, 2N3711.
BC130A	BC108A-183A, 2SC231.
BC130B	BC108B-183B.
BC131	BC108B-109-149-169-173-184-209-239-319-386, GI3711, MPS3711-6521, NKT10519, SK3122, 40399, 2N3711.
BC131B	BC109B-184B.
BC131C	BC109C-184C.
BC132	BC108-148-168-172-183-184-208/A-238A, GI3711, MPS6552, NKT10419, SK3122, SX3710, 40399, 2N3565-5136.
BC134	BC147-167-171-182/B-207/A-237B-385, MPS6575, SK3122, 2N3242-3566.
BC135	BC107/A-147-167-171-118-182A/B-207/A-385, MPS6566-6576, SK3122, 2N3242-3566-3568.
BC136	BC125-140/VI-171-182/A-237/A-211-341VI, GI2270, BSW63, MPS6530-6566, SK3122, 40347, 2N2221A-2194A.
BC137	BC116-126-160/VI-313-327, GI3644, MM3726, MPS6533, SK3114, TIS93, 40410, 2N1132-2906-4030-5042.

BC138	BC140-140/10-160, MM3725, SBC119. SK3024, 2N2219-5189.
BC139	BC140-160/VI, MM3726, MPS6534, SK3025, 40410, 2N1132-2904-3644-4036-5855.
BC140	BC142-144-286, BSY46, MM3725, SK3024, 2SC708, 2N2218/A-2102-2192/A-3036-5321.
BC141	BC142-310, BFX84, BSY55, ME8002, SK3024, 2N2102-2193/A-3020-3036-3569-5262.
BC142	BC141/VI-286, BSY55, ME8003, SK3024, 2N2102-2219/A-2792-3020-3036-3109-5321.
BC143	BC160-161/VI-287-313, BSY46, MM3726, SK3025, 2N1132-4031-4036-4037.
BC144	BC140/X-211-286-341X, BFY50, BSW63, SK3024, 2N2193-2218A-2788-3253-5321.
BC145	BC117-141, BF117-145-257-297, BFR86, MM3001, SE7002, 40373, 2N3114.
BC147	BC171-182/L-207-317, MPS6566, SK3020.
BC147A	BC107A-167A-171A-118-182A/LA-207A-385A, BCY59A, MPS6566-6575, SK3020, 2SC281, 2N3242-3566-3568.
BC147B	BC107B-167B-171B-182B/LB-207B, BCY59B, MPS6566-6576, SK3020, 2N3242-3566-3568.
BC148	BC172-183/L-208-318, MPS6520, SK3020.
BC148A	BC108A-118-168A-172A-183A/LA-208A-386A, BCY58A, GI3710, MPS3710-6520, SK3020, 40450, 2SC281, 2N3568-3710.
BC148B	BC108B/C-118-168B-172B-183B/LB-208B/C-386B/C, BCY58.B/C, GI3710/1, MPS3710-3711-6520, SK3020, 40399-40450, 2N3568-3710-3711.
BC148C	BC108C-118-168C-172C-183C/LC-208C, MPS6520, SK3020, 2N3568.
BC149	BC184/L-209-173-319, MPS6521, SK3020.
BC149B	BC109B-169B-173B-184B/LB-209B-386B, BCY58B, GI3710, MPS3710-6521, SK3020, 40450, 2SC258, 2N3710.
BC149C	BC109C-169C-173C-184C/LC-209C-386C, BCY58C, GI3711, MPS3711-6521, 40399, SK3020, 2N3711.
BC153	BC154-157-159-177-179-204VI-206-212/A-214A, 251-253-259-263-307A/VI, BCY79A, MPS6516-6534, NKT20339, SK3114, 40419, 2N3964-4059-5041.
BC154	BC158-204B-212B-213-214/B-252-307A/B, BCY79B, MPS6522, SK3114, NKT20339, 40419, 2N4059-4359-5087.
BC155	BC121-146, SK3039.
BC156	BC112-121-146, SK3039.
BC157A	BC204A-212LA-251A-291D, 40406, 2N3244-3906-4121.
BC157B	BC204B-212LB-251B-291D, 40406, 2N3245-3906-4121.
BC157VI	BC212L.
BC158	BC178-153-205-213/L, 252-258-321, MPS6522, SK3118, 2SA565.
BC158A	BC204A-205A-213LA-214A-252A, MPS6534, SK3118, 40634, 2N3644-4359.

BC158B	BC204B-205B-213LB-214B-252B, MPS6534, SK3114, 40419, 2N3962-4059.
BC158C	BC213LC.
BC159	BC153-179-206-214/L-253-259-263-322, MPS6523, SK3114, 2SA565.
BC159A	BC214LA, SK3114.
BC159B	BC214LB, SK3114.
BC159C	BC214LC.
BC160	BC143-287, MM3726, SK3025, 2N1132-2904-4037.
BC160-6/10/16	SK3025.
BC161	BC311, SK3025, 2N2904-4036-4037.
BC161-6/10/16	SK3025.
BC167	BC167A-171/A-182/A/L-207-237-317, MPS6566, SK3122, 2SC458.
BC167A	BC107A-118-147A-171A-182A/LA-207A, MPS6566.
BC167B	BC107B-118-147B-171B-182B/LB-207B, MPS6566.
BC168	BC172-183/L-208-238-318, MPS6520, SK3020.
BC168A	BC108A-118-148A-172A-183A/LA-208A, MPS6520, SK3020, 2SC458.
BC168B	BC108B-118-148B-172B-183B/LB-208B, MPS6520, SK3020.
BC168C	BC108C-118-148C-172C-183C/LC-208C, MPS6520.
BC169	BC173-184/L-209-239-319, MPS6521, SK3020.
BC169B	BC109B-149B-173B-184B/LB-209B, MPS6521, SK3020.
BC169C	BC109C-149C-173C-184C/LC-209C, MPS6521, SK3018.
BC170	BC108-148-168-113-183-208-318, MPS6520, SK3122, SX3708.
BC170A	BC148-168-172-183-208-238A, SK3122, SX3710, 2SC458.
BC170B	BC148-168-172-183-209-238A, SK3122, SX3709.
BC170C	BC148-168-172-183-208-238, SK3122, SX3711.
BC171	BC107-113-147-167-182-207-237-317, MPS6566, SK3122, 2SC458, 2N2921.
BC171A	BC107A-182A-237A, SK3122.
BC171B	BC107B-182B-237B, SK3122.
BC172	BC108-114-148-168-183-208-238-318, MPS6520, 2SC458.
BC172A	BC108A-183A-238A, SK3122.
BC172B	BC108B-183B-238B, SK3122.
BC172C	BC108C-183C-238C, SK3122.
B173	BC109-149-169-184-209-239-319, MPS6521, SK3020, 2SC458.
BC173B	BC109B-184B-239B, SK3020.
BC173C	DO109O-104C-209C, 3K3122.
BC174	BC110-182-108, BFR40, SK3024.
BC174A	BC190A, BFR40, SK3024.
BC174B	BC190B, BFR40, SK3024.
BC177	BC157-153-204-212-251-257-261-320, MPS6516/7, SK3114, 2SA565.

BC177A/B	SK3114.
BC178	BC153-158-205-213-252-258-262-321, MPS6518-6522, SK3114, 2SA565.
BC178A/VI	SK3114.
BC178B	SK3053.
BC179	BC154-159-206-214-253-259-263-322, MPS6519-6523, SK3114.
BC179A/B	SK3114.
BC181	BC153-157-177-204-212-251-257-261-307VI-308VI, MPS6519-6523.
BC182	BC107-147-151-167-171/A-174-207/A-237/A-317, MPS3704-6566, SK3122, 2N3242-3566.
BC182A	BC171B-207B-237B, MPS3704, 2N3242-3566.
BC183	BC107-147-167-172-207-208-237/A-238-318, MPS6520, SK3122.
BC183A	BC171A-183B-207A-237A/B-327, MPS3704, 2N3242-3566.
BC183B	BC171B-207B-237B-327, MPS3704, 2N3242-3566.
BC183C	BC207, BCY59D, MPS6575, SE6001, 40450, 2N2926.
BC184	BC109-149-169-173-209-237-239/B-319, MPS6521, SK3122.
BC184B	BC171B-207B-237B, MPS6575, 40450, 2N2926-3566.
BC184C	BC207, BCY59D, MPS6575, SE6002, 40399, 2N2926.
BC185	BC140/X-177-181-211-261, BFY67, FT3722, 40314, 2N1613-1711-2192-2219.
BC186	BC107A-177VI-178-181-261, SK3122.
BC187	BC177VI-178A-179-181-261, SK3114.
BC190A	BC107A-174A, BCY65E/7, BFR16, 2N956-2219A-2484-2645-3904-4409.
BC190B	BC107A-174B, BCY65E, BFR17, 2N956-2219A-2484-2645-3904-4409.
BC192	BC328, BFW20, ME0402, SK3114, 2N2906/7-3485-3644-3720-5354.
BC194	BC246R.
BC196/IV	BC200R.
BC196A/B	BC200G, SK3114.
BC197	BC107-147-167-171-182-207-237B-317, MPS6566, SK3122.
BC197A	BC107-123Y-147-171-182-207, SK3122.
BC198A	BC108-122Y-148-172-183-208.
BC199	BC109-146G-149-169-173-184-209-319, MPS6521, SK3122.
BC199B	BC109-122G-149-173-184-209.
BC200	BC206-322, SK3118.
BC201	BC109-168-200-239, SK3114, 2SA565.
BC202	BC108-200Y-238/9, SK3114, 2SA565.
BC203	BC107-200Y-237, SK3114, 2SA565.
BC204	BC157-177-153-212-251-257-261-307, MPS6516/7, SK3114.
BC204VI	BC177VI-212-251A-291D-307VI, MPS3703, SK3114, 40406. 2N3133-3251.

BC204A	BC177A-212A-251B-291A-307A, MPS3702, SK3114, 40406, 2N3133-3251.
BC204B	BC177B-212B-251C-291D-307B, MPS3702, SK3114, 40419, 2N3134-3251.
BC205	BC153-158-178-213-252-258-262-308, MPS6518-6522, SK3114.
BC205VI	BC116-178VI-213A-252A-308VI, EN3502, MPS6519, SK3114, 40419, 2N3134.
BC205A	BC116-178A-205A-252B-308A, EN3502, MPS6519, SK3114, 40419, 2N213B-3702.
BC205B	BC116-178B-213B-252C-308B, EN3502, MPS6519, SK3114, 40419, 2N3134.
BC206	BC154-159-179-206-214-253-259-263-309, MPS6519-6523; SK3114.
BC206B	BC179B-214B-263B-291D-309B, SK3114, 40419, 2N3251-3644.
BC207	BC107-113-147-167-171-182, MPS6577, SK3122.
BC207A/B	BC171B-182B-237B-385B, MPS6575, SK3122, 2N3242-3566.
BC208	BC108-148-168-172-183, MPS6520, SK3122.
BC208A	BC172A-183A-238A-386A, MPS6574, SK3122, 40450, 2N2924-5136.
BC208B	BC172B-183B-238B-386B, MPS6574, SK3122, 40450, 2N2925-5136.
BC208C	BC172C-183C-238C-386C, MPS6574, SK3122, 40399, 2N2926-5136.
BC209	BC109/A-149-169-173-184, MPS6521, SK3122.
BC209B/C	BC174C-184C-239C-386C, MPS6574, SK3122, 40399, 2N2926-5136.
BC211	BC140/VI-120, BFR17, BSX45VI, SK3024, 40290, 2N2102-2218/A-2219A-2193A-3020-3036-3567-3725-4409.
BC212	BC153-157-177-204-251-257-261-307-320, MPS6516, SK3114.
BC212A	BC204-266A-307A-327, ME0411, 40314, 2N3790-3906-4121.
BC212B	BC204-266B-307B-327, ME0412, 40314, 2N3790-3906-4121.
BC213	BC153-157-177-204-251-257-261-205-307/VI-321, MPS6522, SK3114.
BC213A	BC204-266A-307A-327, ME0411, 40314, 2N3790-3906-4121.
BC213B/C	BC204-266B-307B-327, ME0412, 40419, 2N379-3906-4121.
BC214	BC154-159-179-253-259-263-206-307/A-322, MPS6523, SK3114:
BC214A/B/C	BC204-266B-307B 327, ME0412, 40419, 2N3790-3906-3962-4059-4121-4249.
BC215	BC149-157-231-261-327, MPS6534.
BC215A	BC153-160VI-290-313-327, 2N4031-4236-5323.
BC215B	BC154-160X-290-313-327, 2N2907-3799-403B-4236-5323.

BC216	BC107A, SK3024.
BC219A	2N4360.
BC219B	2N4343.
BC219C	2N4342.
BC220	BC113-170C-209B-237A-239B/C-386B, MPS6544, SK3122, SX3710, 40399, 2N2926-5136.
BC221	BC328, BFY64, BSV15X, BSX40, ME0402, SE8540, SK3114, NKT20339, 40406, 2N2906-5226-5365.
BC222	BC125-338, BSX22-70-75, NKT10339, ME8061, SE8040, 40407, 2N2221-3705-4424-5225.
BC223	BC107-120-140/XVI-144-147-130-207-223-337-340XVI, MPS3705, BFX55, BFY68, SE8002, 2N2218-2788-3020-3553-3704-4046-4952.
BC224	BC153-158-162-178-204B-205-213-225-252-258/B-262-307B-308B, BCY78C, MPS6518-6522, NKT20339, 2N3136-4037-5244.
BC225	BC204A-212-224-251A-307A/VI, MPS6518, NKT20339, SK3118, 2N3136-4037-5244.
BC231	BC137-160-161-177-179-224-225-257A-261-327, BSW45A, FPS6535, NKT20339, 40406, 2N3251-5244.
BC231A/B	BC327.
BC232	BC107-130-136-140-147-167A-207-225-231-337-340, BSW43A, FPS6532, NKT20339, 40406, 2N2222-3251.
BC232A/B	BC337.
BC237	BC107-147-167-171-113-182-207-317, MPS6566, SK3122.
BC237A	BC107A-144-182A-171A-207A-232, MPS3705, NKT10339, 40407, SE8002, 2N2788.
BC237B	BC107A/B-171B-207A/B-385A-182B, MPS6575, 40450, 2N3566.
BC238	BC108-114-148-168-172-183-208-318, MPS6520, SK3122.
BC238A	BC107B-108A-183A-207B-172A-208A, MPS6575, 40450, 2N3566.
BC238B	BC108A/B-172-183B-208A/B-386A, MPS6574, 40450, 2N2924-5136.
BC238C	BC108B/C-172C-183C-208B/C-386B, MPS6574, 40450, 2N2925-5136.
BC239	BC109-149-169-173-184-209-319, MPS6521, SK3122.
BC239B	BC108C-109B-173B-184B-208C-209B-386C, MPS6574, 40399, 2N2926-5136.
BC239C	BC109B/C-173C-184C-209B/C-386B, MPS6574, 40399, 2N2925-5136.
BC250	BC153-177-205-231-308-321, MPS6516, SK3114, SX4059.
BC250A	BC109C-178A-205VI-209C-258A-260A-308A/VI-386C, MPS6574, SK3114, 40399, 2N2926-5136.
BC250B/C	BC205A/VI-178B-212-258B-260B-307VI-308B, MPS3703, SK3114, 40406, 2N3133-3251.
BC251	BC153-157-177-204-212-257-261-307-320, MPS6516, SK3114.

BC251A	BC204VI-205B-212A/B-261A-307A/B/VI, BCY79A, MPS3702, SK3114, 40419, 2N3134-3251:
BC251B	BC204VI-204A-208B-213A-212B-261B-307A/B, BCY79B, SK3114, 40314, 2N3790-3906-4121.
BC251C	BC204A/B-212B-213B-261C-307B/C-308A/B, SK3114, 40314, 2N3790-3906-4121.
BC252	BC153-158-178-205-213-258-262-308-321, MPS6522, SK3114.
BC252A/B	BC204B-205VI-213A/C-262A-308A/B/VI, BCY78A, SK3114, 40419, 2N3790-3906-4121.
BC252C	BC205A/B-213C-214A-262C-308B/C-309A, BCY78C, FT3645, MPS6519, SK3114, 40419, 2N3644.
BC253	BC153-159-179-206-214-259-263-309-322, MPS6523, SK3114, 2SA565.
BC253A/B	BC206B-214A-309A/B, BCY78B, FT3645, SK3114-40419, 2N3133-3251.
BC253C	BC263C-309B/C, SK3114.
BC254	BC108A-110-107-130-145-147-174-206B-207-214C-307A-309B, BSY55, FT3645, 40419, 2N3134-3251.
BC255	BC107-108B-130-145-147-337, BF178, BFY43.
BC256	BFR80, SK3114, 2N4250A.
BC256A	BC117-161VI-254-266A-307A, BCY79A, BSW44A, ME0411, SK3114, 2N871-1890-1893-5320.
BC256B/C	BC161VI-204-256A-266B-307A-BCY79A/B/C, BSW44A, MPS6534, SK3114, 40634, 2N3644-4359.
BC257	BC153-157-177-204-212/L-231-251-261-307, MPS6516/7, SK3118, 2SA565.
BC257A/B/VI	BC177C-204B-224A-251C-287A-308A/B, FT3645, MPS6516/7, 40419, 2N3644.
BC258	BC153-158-178-205-213/L-251/2-262-307, MPS6518-6522, SK3114, 2SA565.
BC258A/B/C	BC178C-205B/C-225C-252C-291D-308C, MPS6519, 40419, 2N4059-5244.
BC259	BC153-159-179-206-214/L-253-263-309, MPS6519-6523, SK3114.
BC259A	BC212LA.
BC259B	BC179B-206B-214LB-253B-283B-309B, MPS6522, 40406, 2N2696-3134.
BC259C	BC179C-206C-214C-253C-283-309C, MPS6522, 40406, 2N2696-3134.
BC260	SK3114, 2N3964.
BC260A	BC178A/VI-204A-212-250A, MPS6519, SK3114, 40406, 2N3133-4121.
BC260B	BC178A/VI-204B-212B-250B, MPS6519, SK3114, 40406, 2N3134-4121.
BC260C	BC178B/C-204-212C-250C, MPS6519, SK3114, 40419, 2N3134-4121.
BC261	BC153-157-177-204-212-251-257, MPS6516, SK3025, 2N3964.
BC261A	BC177A/VI-204III-212-251A-307A, SK3025, 40406, 2N3133-3906-4121.
BC261B	BC177A-204A-212A-251B-307B, BCY79IX, SK3025, 40406, 2N3133-3906-4121.

BC261C	BC177B-204B-212B, 251C-307C, SK3114, 40406, 2N3134-3906-4121.
BC262	BC153-158-178-205, 213-252-258-262, MPS6522, SK3114, 2N3964.
BC262A	BC178A/VI-205VI-213-252A-308A, SK3114, 40314, 2N3251-3790.
BC262B/C	BC178B-205B-213B-252C-308C, MPS3702, SK3114, 40314, 2N3251-3790.
BC263	BC153-159-179-206-214-253-259, MPS6523, SK3114, 2N3964.
BC263A	BC179A/VI-206VI-214-253A-309A, FT3645, MPS6516, SK3114, 40419, 2N3133.
BC263B/C	BC179B/C-206B/C-214B/C-253B/C-309B/C, FT3645, MPS6517/8, SK3114, 40419, 2N3134.
BC266	BC143-212, SK3114, 2N3962-3965.
BC266A	BC139-161/VI-256A-307A, BFY79A, BSV16VI, BSW44A, ME0411, SK3114, 2N2906A-3250A-4037.
BC266B/C	BC139-161X-166VI-256B-307B, BCY79B/C, BSV16VI/X-BSW44A-45A, ME0411/2, SK3114, 2N2906A/7A-3250A/1A-4037.
BC267	BC140-144-207-211-317, SK3122, 2N2102-2218-3020-3036.
BC267A	BC107A-171A-207A-237A-385A, MPS6575, PBC107A, 2N3242-3566.
BC267B	BC107B-171B-207B-237B-337-385B, MPS6575, PBC107B, 2N3242-3566.
BC268	BC140-208-211-318-338, SK3122, 2N2102-2218-3020-3036.
BC268A	BC108A-172A-208A-238A-386A, PBC108A, MPS6574, 40450, 2N2924-3565.
BC268B/C	BC108C-172C-208C-238C-386C, PBC108C, MPS6574, 40399, 2N2926-3565.
BC269	BC109-209-319, SK3122.
BC269B/C	BC109B/C-173B/C-209B/C-239B/C-386B/C, PBC109B/C, MPS6574, 40399, 2N2925/6-3565.
BC270	BC140-144-223-283-338, SK3122, 2N2218-3020-2102.
BC270A	BC108A-170A-208A-238A-386A, MPS6574, PBC108A, 40450, 2N2924-3565.
BC270B/C	BC108B/C-107B/C-208B/C-238B/C-386B/C, PBC108B/C, 40399, 2N2925-3565.
BC271	BC283-338.
BC271A/B	BC172A, BSX45VI/X, ME8001, NKT10339, 40407, 2N2222-2645-3568-4409.
BC272	BC337, 2N2222.
BC272A/B	BC171A/B. BSX45XVI, ME8001, 40407, 2N2222-2645-3568-4409.
BC280	SK3122, 2N930.
BC281	SK3114, 2N3964.
BC282	BC190A, BCY65VII, BSX45XVI, SK3122, 2N2483-2952-3053-3946-3947.
BC283	BC313-360X, BSV15VI, SK3114, 2N2906/7-4030-4236.
BC284	SK3122, 2N915.
BC284A	BC140X-184A, BCY59A, BSX51A, SK3122, 2N2218-8553-4237.

BC284B	BC140XII-184B, BCY59B, BSX52A, SK3122, 2N2219-3553-4237.
BC285	BF117-178-257-336, SK3122, 40373, 2N2243-3114-3723-4001-4895.
BC286	BC140XVI-141XVI, BFY46-68, BSY54, SK3024, 2N1711-3568-3725A, 40290.
BC287	BC161VI/X-313, SK3025, 2N2904A-4030, 4037-4236.
BC288	BC140X-211, BFX34, BFY55, BSX45X-62, ME8003, SK3024, 2N3109-3110-3553-3567-4047.
BC289	BC107, SK3122.
BC289A	BC171A-207A-382A, BCY59B/D, MPS6576, SK3122, SE6001, 2N3242A-3665.
BC289B	BC171B-207B-382B, BCY59C, MPS6576, SK3122, SE6002, 2N3665.
BC290A	BC107B.
BC290B	BC172B-207B-283B, BCY59C, MPS6574, SK3122, 2N3242A-3665-5136.
BC290C	BC172C-207-383C, BCY59D, MPS6574, SK3122, 2N3242B-3665-5136.
BC291	BCY79VIII, SK3114.
BC291A	BC204A-224A-307A, BCY79B, MPS6519, SK3114, 40419, 2N3790-4121.
BC291D	BC204VI-224B-307VI, BCY79A, MPS6519, SK3114, 40419, 2N3790-4121.
BC292	BCY79X, SK3114.
BC292A	BC204A-214A-307A, BCY79B, MPS6516, SK3114, 40406, 2N3133-4121.
BC292B	BC204B-214B-307B, BCY79C, MPS6517, 40406, 2N3134-4249.
BC293	BFX34, BSV64, BSX45VI-63, 2N1479-1613-3405-3418-3553-3568-5531.
BC297	BC143-160-227, MM3726, 2N1132-4037.
BC297B	BC307A.
BC297PA/B	BC204A/B-214A/B-307A/B, BCY79A/B, MPS6519, 40419, 2N3790-4121.
BC298	BC143-160-328, MM3726, 2N1132-4039.
BC300	BC141-293, BSV64, BSX46X-47, ME8002, 2N1893-2102-3019-3020-3036-3553-3568.
BC301	BC141/VI-211-293, BSW65, BSX23-46X-62, ME8003, MM3001, SK3024, 2N286-2102-2243A-2351-3032-3036-5184.
BC302	BC140/VI-144, 2N2102-2218-3020-3109.
BC303	BC161/VI-287-313, SK3025, 2N4030-4036/7-4236.
BC304	BC143-160/VI-287, MM3726, 2N1132-4037.
BC307	BC153-157-177-204-212-251-261-320, MPS6516/7 SK3114.
BC307VI	BC177VI-204VI-212-214A-261A, MPS6519, 40419, 2N3790-4121.
BC307A	177A-204A-212A-214A-251A-261A/B, MPS6519, 40419, SK3114, 2N3790-4121.
BC307B	BC251B-261B.
BC307C	BC251C-261C.

BC308	BC153-158-178-205-213-252-262-321, MPS6518-6522, SK3114.
BC308VI	BC178A-205VI-214A-213-262A, MPS6519, 40419, 2N3790-4121.
BC308A	BC178B-205A-213A-224A-252A-262A/B, MPS6519, 40419, SK3114, 2N3790-4121.
BC308B	BC178B-205B-213B-252B-262B/C, MPS6519, 40406, 2N3133-4121.
BC308C	BC252C-262C.
BC309	BC153-159-179-206-214-253-263-322, MPS6519-6523, SK3114.
BC309A	BC206A-214A-179A-253A-263A, 40419, SK3114, 2N3790-3906-4121.
BC309B	BC206B-179B-214B-253B-263B, 40419, 2N3790-3906-4121.
BC309C	BC253C-263C.
BC310	BC286.
BC311	BC287.
BC313	BC160VI-161/VI-287-313, BSV16VI, SK3024, 2N4030-4036/7-4236.
BC314	BC293, BD139, BSV64, BSW10, BSX63, 2N1479-2243A-2351-5184-5531.
BC315	BC153-159-177A-179-204A-206-214/B-253-261B-263-415A, MPS6523, 40419, 2N3790-3906-4121-5087.
BC320/A/B	BC327.
BC327	BC161XVI-287, TIS93, 2N2907A-4030-4037-4236-5368.
BC328	BC160XVI-207, TIS93, 2N2907-4030-4037-4236-5368.
BC329	2N5210.
BC333	BC239B.
BC334	BC309A.
BC335	BC239B.
BC336	BC309A.
BC337	BC140XVI-288, TIS92, 2N2222A-3110-3553-3567-3725-4047-4424.
BC338	BC140XVI-288, TIS92, 2N2222-3110-3553-3567-3725A-4047-4424.
BC340	2N3109-3110.
BC340VI	BC140VI-288, SK3024, 2N2218A-3110-3553-3567-3725-4424.
BC340X	BC140X-288, SK3024, 2N2219A-3110-3553-3567-3725-4424.
BC340XVI	BC140XVI-289B, SE6002, SK3024, 2N2219A-3110-3553-3567-3725A-4424.
BC341	2N3107-3108.
BC341VI	BC141VI-289A-383A, MPS6576, SE6001, SK3024, 2N3242A-3417-3665.
BC341X	BC141X-289A-383A, MPS6576, SE6002, SK3024, 2N3242B-3417-3665.
BC341XVI	BC141XVI-289B-383B, MPS6576, SE6002, 2N3242B-3417-3665.
BC342	BC141VI.
BC343	BC161VI.
BC344	BD139.

BC345	BD140.
BC347	BC237A/B.
BC348	BC237A/B.
BC349	BC308VI/A.
BC350/1	BC307VI/A.
BC352	BC308VI/A.
BC354	BC238A/B.
BC355	BC308VI/B.
BC357	BC308.
BC360	BC143-160, MM3726, 2N1132-4037-5022-5023.
BC360VI	BC160VI-287, SK3025, 2N2904/A-4030-4037-4230.
BC360X	BC160X-287, SK3025, 2N2905/A-3644-4032-4033-4037.
BC360XVI	BC160XVI-287, SK3025, 2N2905/A-3644-4032-4033-4637.
BC361	BC143-157-160-161-177, MM3726, 2N1132-4030/1/2-4037.
BC361VI	SK3025, 2N2904A.
BC361X	SK3025, 2N2905A.
BC370	BC143-160-328, MM3726, 2N1132-4037.
BC377	BC140-144-337, 2N2102-2218-3020-3036.
BC378	BC140-144-338, 2N2102-2218-3020-3036.
BC381	BC177/A-204A-225-261/A-307VI-328, EN3250, 2N3250-3702-4403.
BC382	BC107-113-147-167-171-182-207-414A/B, MPS6566.
BC382B/C	BC207-237A/B-266A/B, MPS6575-40450, 2N3566.
BC383	BC108-114-148-168-172-183-208-237-413B/C, MPS6520.
BC383B/C	BC207-237C-266C, MPS6575, 40450, 2N3566.
BC384	BC109-149-169-173-184-209-413B, MPS6521.
BC384B/C	BC207-237B-261B-337, MPS6575, 40450, 2N3566.
BC385	BC107-113-147-167-171-182-207-237/A-261B-337, MPS6566-6575, 40450.
BC386	BC108-114-148-168-172-183-208-238/B-262C-338, MPS6520-6575, 40450, 2N3566.
BC387/8	BC337.
BC397	BC161VI.
BC398	BC141VI.
BC399A	BC146 Yellow.
BC399B	BC146 Green.
BC407	BC237.
BC408	BC238.
BC409	BC239.
BCW10	ZXT300.
BCW11	ZXT500.
BCW12	ZXT301.
BCW13	ZXT501.
BCW14	ZXT302.
BCW15	ZXT502.

BCW16	ZXT303.
BCW17	ZXT503.
BCW18	ZXT304.
BCW19	ZXT504.
BCW20	ZXT330.
BCW21	ZXT530.
BCW22	ZXT331.
BCW23	ZXT531.
BCW29	BC153-158A-204A-308A, BCY78A, GI3702, MPS3702, SK3118, 40319, 2N978-4060.
BCW30	BC153-158B-204B-308B, BCY78B, GI3702, MPS3702, SK3118, 40319, 2N2411-4060.
BCW31	BC132-148A-183-208A, BCY58A, GI3709, SK3039, 2N930-2483-4135.
BCW32	BC148B-183C-208B-284, BCY58B, GI3710, SK3039, 2N930-2484-5219.
BCW33	BC108C-148C-208C-384C, BCW32, BCY58C, FT107C, GI3711, 2N2484-5223.
BCW34-36	BC140X-144-211-340X, ME8003, SK3122, 2N2222A-2484.
BCW37	BC143-160X-313-360X, ME0402; SK3114, 2N2907A-5323.
BCW46	BC140XVI-211, BCY86, BSX45XVI, SK3024, 2N2222A-2483-3568.
BCW47	BC147B-182A-204A-207/B-317, ·BCY58B, MPS6576, SK3024, 2N2484.
BCW48	BC148C-183-204C-208/C-318, BCY58C, MPS6575, SK3024, 2N2484-3569.
BCW49	BC149C-184-205C-209/C-319, BCY59C, MPS6571, SE4002, SK3024, 2N2483-3569.
BCW56	BC161X-287-313, BSV16X, BCW85, SK3118, 2N2907A-4249-4356-5323.
BCW57	BC157A-204-207A-291A-320, BCW86, BCY58A, SK3118, 2N3486-4143-5323.
BCW58	BC158B-205-208B-292-321, BCY59B, GI3702, MPS3702, SK3118, 2N4037-4058-4359.
BCW59	BC159B-206-208C-292-322, BCY59C, GI3703, MPS3703, SK3118, 2N4037-4058-4359.
BCW85	BC139-313-361X, 2N2907A-4036-4356.
BCW86	BC177A-313-361X, BSV16X, 2N2907A-4036-4356.
BCX10	2N1132.
BCY10	BCY38, ME503, MPS404A, SK3114, 2N329B-2945-3677.
BCY11	BCY39, ME513, SK3114, 2N329A-1232-3062.
BCY12	BC143-160, BCY40, BSW20VI, BSY59, NKT20339, SE8540, SK3114, 2N721-1131/2.
BCY13	BC144, BFX61, BFY12-40-44-50, ME8003, SK3122, 40347, 2N497-656-2194-2219-2410.
BCY14	BC142, BFX65, BFY13-41-65, BSY55, ME8002, 2N657-698-1890-5320.
BCY15	BC144, BFX61, BFY40-44-13-51, ME8003, SE8001, SK3122, 40347, 2N656-2194-2219-2410-4140.

112

BCY16	BC142, BFX84, BFY41-65, BSY55, ME8002, SK3122, 2N657-698-1890-2219-5320.
BCY17	BCY27-38, BSX36, MPS6535, SE8540, SK3025, TIS38, 2N3644-5243.
BCY18	BCY28-38, BSX36, MPS6535, SE8540, SK3025, TIS38, 2N3644-5243.
BCY19	BCY40, BFX74A, SK3025, 2N2605-3072-3250A-3468-3962.
BCY20	MM4000, 2N3495-3931-5322.
BCY22	BCY29-39, BFS91, SK3025, 2N3496-4356-5323.
BCY23	BCY29-40, BSX36, MPS6535, SE8540, SK3025, TIS38, 2N3644.
BCY24	BCY29-38, BSX36, MPS6535, SE8540, SK3025, TIS38, 2N3644.
BCY25	BCY29-54, MPS6535, SE8540, SK3025, TIS38, 2N3644.
BCY26	BCY29-38, BSX36, MPS6535, SK3025, TIS38, 2N3644-5243.
BCY27/28	BC40, BSX36, MPS6580, SK3025, 2N3638-3829-5040.
BCY29	BCY34, BSX36, MPS6580, SK3114, 2N3638-3829-5040.
BCY30/31/32	BCW56A, BCY39, BFX74A, SK3025, 2N3250A-3496-3648-5323.
BCY33	BCY40-78VII, BSX36, MPS6580, SK3025, 2N3638-3829-5040.
BCY34	BC177-212-216-261, BCY27-29-38-78VII, BSX36, MPS6535, SK3025, TIS38, 2N3644-5040.
BCY38	BC143, BCY29, MPS6535, SE8540, SK3114, 2N2904-3644-3762.
BCY39	BC160-161, BFX74A, SK3114, 2N2904/A-3072-3496-3763-4027-5323.
BCY40	BC143, BCY29, MPS6535, SE8540, SK3114, 2N2904-3644-3762.
BCY42	BFY39I, BSW41, BSX25-26-48, SK3122, SE6001, 2N708-3935-4123-4148-4876.
BCY43	BFY33-39I, BSW41, BSX25-26, SE6001, SK3122, 2N708-3933-4123-4148-4876.
BCY50	BC108A/B-183/B, BSX68, BSY62, ME3001, SK3122, 2N709-743-4251-5186-5220.
BCY50R	BC109B-184B
BCY50I	BC108B-130B-183B-SK3122
BCY51	BC108A/B-107/B-182B-183/B, BSX26-68, BSY62, ME3002, SK3122, 2N744-2432-3564-3600-4264-5137
BCY51I	BC107B-129B-182B, SK3122
BCY51R	BC109B-184B, MPS6521
BCY54	BC160VI-313-360VI, BSV16VI, ME0401, SK3114, 2N2904A-3120-3121-3468
BCY55	2N2060B
BCY56	FT107B, SE4021, SK3122, 2N2484
BCY56A	BC207-382A, BCY59A, GI3705, MPS6575, SE1001, 40235
BCY56B	BC207-382B, BCY59B, GI3694, MPS6575, SE1002, 40237

BCY57	BC107A-182A, BCY58, FT107A, SE4022, SK3122, 2SC648, 2N2484-2921-3568
BCY57A/B	BC208-383B, BCY58B/C, GI3706, MPS3706, SE2001, 40450
BCY57C	BC208-383C, BCY58C, GI3706, MPS3706, SE4002, 40399
BCY58	BC107B-182B, AM261, BCY38, BFY44, FT107A, MPS6566, SE4022, SK3020, 2SC907, 2N3568
BCY58A/B	BC208-383B, BCY58VII/VIII, GI3708, MPS3708, SE2001-5025, SK3122, 40450
BCY58C	BC208-383C, BCY58IX, GI3708, SE4010, SK3122, 2N5088-5135
BCY58D	BC208-209-383C, BCY58X, GI3708, SE4010, SK3122, 2N5088-5135
BCY59	BC107B-182B, AM262, FT107B, MPS6566, SE4021, SK3020, 2SC907
BCY59A	BC207, BCY59VII, MPS3693, SE1001, SK3122, TIS97, 40407, 2N3693
BCY59B	BC207, BCY59VIII, MPS3694, SE1002, SK3122, TIS97, 40407, 2N3694
BCY59C	BC207-382C, BCY59IX, MPS6575, SE4021, SK3122, 2N3694-5106
BCY59D	BCY59X, SK3122
BCY65	BC107B-174, FT107C, SE4020, SK3024, 2SC648, 2N2483/4
BCY65A	BCY86, BFR16, BSX52A, 2N2483
BCY65EVIII	BCY59B-86, BSX52A, BFR17, 2N2484
BCY66	BC107/B-109-214, BCY59VIII, FT107B, MPS6566, SE4021, SK3024, 2SC648, 2N2484
BCY66VII	BC207-289A/B, BCY59B/C, MPS6575, 2N929-930-3694
BCY67	BC325, BCY71, SK3025
BCY67VII	BC204-292A, BCY79B, 2N3798-3964-5086
BCY67VIII	BC204-292B, BCY67VIII-79C, 2N3799-3964-5087
BCY70	BC161VI-177-204-212-216-261-287-361VI, BCY79, SK3114, 2SA549, 2N2906A-3250-3964
BCY71	BC160VI-204-292A-360VI, BCY79B, SK3114, TIS93, 2N3644-3964-4359
BCY72	BC160VI-116-205-360VI, BCY78B, MPS3638, SK3114, 2N2696-3638-3644-3964
BCY78	BC178, SK3114, 2N3964
BCY78A	BC206-283, SK3114, 2N3964-4059-5138
BCY78B/C/D	BC205-283, 2N3964-4059-5138
BCY79	BC179, SK3114, 2N3964
BCY79A/B/C	BC204-292A, SK3114, 2N3798-3964-5086
BCY85	BC141XVI, BFR21, BSY56, 2N1890-2405-3499-4410
BCY85A	BCW46A
BCY85B	BCW46B
BCY86	BC141XVI, BFR20, BSY54, 2N2511-4409
BCY86A	BC107B
BCY86B	BCY59X
BCY90/1/2	BC160VI-287, BCY19-54, GI3644, ME0411, MPS6516, SK3114, 2N3644-4121-5365-5448
BCY93	BCY22-30, BFX40, SK3114, 2N2906-3494-4027-4031

BCY94/95	BCY22-31, BFX40, SK3114, 2N2906-3494-3496-4027-4031-5323
BCY96/7	BCY20, BFX40, MM4000, SK3025, 2N4036-4037-4928
BCZ10	BC153-177-178-204-206-212-213-216-261-360VI, BCY72-78B, BSV15VI, BFX48, MPS6522, GI3638, SK3114, 2N2695-3638-3829
BCZ11	BC177-204-206-212-216-261, BCY72-78A, BSX36, GI3638, MPS3638, SK3114, 2N2695-3638-3829
BCZ12	BC360VI, BCY70, BFX40, BSV16VI, ME0411, SK3114, 2N2905-3763-4026-4030-4037-5365
BCZ13	BC153-178-200-205-206-213-261-262, BCY27-72-78A, BCZ33, BSX36, GI3638, MPS3638-6522, SK3114, 2N2695/6-3638-3829
BCZ14	BC153-178-200R-205-206-213-261-262. BCY28-72-78A, BSX36, GI3638, MPS3638-6522, SK3114, 2N2695/6-3638-3829
BD106	BD109-124, BCY48A, BDY34, BUY24, MJ481, SE3035, SK3054, 25D70, 2N3054-3441-4910
BD106A	BD106-109C-124, BCY47A, BDY34, SK3054, 2SC830, 2SD71, 2N3054
BD107	BD124, BDY13X-34, BUY34, MJ481, SE3035, SK3054, 25D143, 2N3054-3441-4911-4912
BD107A	BD107-124, BCY474, BDY13X-34, 25D71, SK3054, 2SC830, 2N3054
BD107B	BD107, BDY13XVI, SK3054, 2SC830, 2SD71
BD109	BD107-116-124, BDY34-39, MJ2249, SK3054, TIP31A, 2SD71-143, 2SC830, 2N1484-3054/5
BD111	BD107-124-130, BDY34-39-72-92, SK3027, 2SC681A, 2SD71-144, 2N3054/5
BD112	BD124-129, BDY27-39, SK3027, 2SC830, 2N1484-3055
BD113	BD145, BDY15-34-39, SK3027, 2SD144, 2N1484-3055
BD115	BD120-127-129, BDY56. BF115-179C-258-279C, SFC258, SK3045, 2SC154G, 2N3847
BD116	BD117-127-130-145, BDY13-16-34-71, SK3027, 2N2821/2-3055
BD117	BLY47, BD123-127-130, BDY11-13X-34-39-55, SK3017A, 2SC1030, 2N1490-1618-2820-3055-3233
BD118	BLY49, BD123-127-130-145, BDY34-55, SK3027, 2N2822-3233
BD119	BD127-129-144, BDY29, BLY50, BU111, 40328, 2N2580-3902
BD120	BD144, BDY74, BU110-111, 2N2580-2821-4301-5038
BD121	BD107-109-111-113-124-130, BDY34, SK3027, 2SD71, 2N1060-1016-3054/b
BD123	BD130, BDY20-34, TIP41B, 2N1212-1488-1618-3054
BD124	BD107-109-116-131, BDY13, MJ2249, SK3054, TIP31A-41B, 2N1209-1487-1618-3054/5

BD127	BD115, BF118-258, BU111, MJ3440, SK3017A, 2N5345-5415/6
BD128	BD144, BF259, BU111, BUY62, MJ3439, DTS413, 2N3079-3434-3439-5416
BD130	BD120, BDY19, PP3001, SK3027, TIP4/B, 40363, SC1030, 2N3055-3235-3667-3771
BD131	BD118, BDY13-16, MJE205, SK3054, TIP33-41A, 2N2822-3055
BD132	MJE101, SK3083, TIP32A
BD135	BD137, MJE202-521, SK3054, TIP29, 40635, SDA345 2N3053-3055
BD136	MJE102-371, SK3083, SDA445, TIP30, 40634, 2N4037
BD137	MJE203, SDA345, SK3054, TIP29A, 40635, 2N2102-4922
BD138	MJE103, SDB445, SK3083, TIP30A, 40634, 2N4036-4919
BD139	MJE204, SD345, SK3054, TIP29B, 40594
BD140	MJE104, SD445, SK3083, TIP30B, 40595 2N4036-4920
BD141	BD120-129-130, BDY19-58, 2N2580/1-3055-3442-4347-5038-5303
BD142	BD130, BDY73, BLY17-72, SK3027, 2N3055
BD144	BD119-128, BDY74, BUY60, 2N2580-3079-3439
BD145	BD113-130, BDY34-75, BLY47, SK3027, 2N3080-3232-3252-3863/4
BD162	BD106-109-124-131, MJ3101, SK3054, TIP31, 2N3054-4910
BD163	BD107-109-124-131, MJ2249, SK3054, TIP31A, 2N3054-4910
BD165	BD131
BD166	BD132
BD167	BFX34
BD168	BFS92
BD170	2N4033
BD171/2	BSW67
BD173	BD137
BD175	BD131
BD176	BD132
BD177	BDY61
BD179	BU126
BD185	BD131
BD186	BD132
BD187	BD131
BD188	BD132
BD189	BDY61
BDX10	2N3055
BDX11	2N3442
BDX12	2N4347
BDX13	BDY38, 40251
BDX23	40636
BDX24	40250
BDX40	2N3772

BDX41	2N3771
BDX50	2N3773
BDX51	2N4348
DBX60	2N3055U
BDX61	2N3055V
BDX70	2N6098
BDX71	2N6099
BDX72	2N6100
BDX73	2N6101
BDX74	2N6102
BDX75	2N6103
BDY10	BDY24-38, BLY72, BUY14, SK3036, 2N3055-3863-4907
BDY11	BD124, BDY34-39, BLY72, SE3035, SK3036, 2N3055-4909
BDY12	BD124, BDY23-34, BLY47A-66, MJE202, SK3054, 2SC830H, 2SD71, 2N3054-3578-4300-4910
BDY13	BD124, BDY24-34, MJE202, SK3054, 2SC830H, 2N3050-3054-3055-3878-4301
BD15/A/B/C	BD124, BDY12-34-180A, BLY47/A-74, 2SC830H, 2SD71, SE3035, 2N3054-3441-4910-5428
BDY16/A/B	BD124, BDY13-34-81, BLY48-74, SE3035, 2N3055-3441-4911/2-5429
BDY17	BD130, BDY73, BLY49-72, SK3027, 2N2819-3055-3445-3771
BDY18	BD130-183, BDY72-74, BLY50, 2N2820-3055-3446-3772
BDY19	BD130, BDY73-74, BLY49, 2N2821-3055-3442-3446-3773
BDY20	BD130, BDY74, BLY50-72, 2N3055
BDY23	BDY18-92, BLY49-70, BUY13, SK3027, 2N3055-4914
BDY24	BDY91, BLY50-66, BUY13, 2N3055-4915
BDY25	BUY18-20-26, MJ423, 2N2822-3442-3583-4347-5264
BDY26	BD110, BUY18-21, MJ423, 2N3079-3584-5264-5387
BDY27	BU126, BUY22-28, MJ423, 2N3080-3585-5388
BDY28	BU126, BUY23, MJ423, 2N3080-4240-5389
BDY34	BD124-137, BDY17-71, BUY19-20, SK3054, 40250, 2N1483-3054/5-3766-5152
BDY38	BD130-138, BDY73, BUY20, BLY72, SK3027, 2SC665H, 2N3252-3055-3771-4914-5006
BDY39	SK3027, 2N3055
BDY49	BDY19
BDY53	BD130, BDY92, BLY17, TIP40, 2N3055-3235-5672
BDY54	BUY18-26, PP3001, 2N3235-3442-3846-5264
BDY55	BD130, BDY20, 2N456A-3055-3235-5288
BDY56	BDY19, BUY26, MJ423, 2N3233-3583-4347-5264-5387
BDY57	BDY19, BUY12, 2N2772-3232-3716-3772-5288-5403
BDY58	BDY19, BUY20-26, 2N2821-3773-4348
BDY60	BDY13-57, BUY13, 2N1486-3055-3232-3441-3716-5288

BDY61	BD130, BDY13-53, 40543, 2N3055-3551-3713-3716
BDY62	BD109, BDY13-34-81, SK3009, TIP33A, 2N1485-3551/2-5295
BDY71	BDY18-61, BLY47, BUY13, 2N3054/5-5297
BDY72	BDY19, BUY26, 2N1022-1618-3233-3441-3551-4347-5284
BDY73	BD130, BDY20, 2N3055-3235-3715-5038
BDY74	BDY19, BUY12, 2N3442-3647-3773-4003-4347-5039-5289
BDY75	BDY19, 2N2819-3771-5302
BDY76	BDY19, 2N2819-3055-3772
BDY77	BDY19, BUY26, 40411, 2N2773-2821-3236-3442-3773-3846
BDY78	BDY18, BUY13, BLY47, 2N1490-3054/5
BDY79	BDY19, BUY26, 2N1022-3233-3441-3551-3667-4347-5284
BDY80	BD106, BDY34-38-12, TIP33, 2N1069-1485-3232-5002
BDY80A	BD131
BDY81	BD107, BDY13-17-34, TIP33A, 2N1070-1486-3232-5004-5293
BDY81A	BD131
BDY82	TIP34, 2N1546-4398-5003
BDY82A	BD132
BDY83	BDY82, TIP34A, 2N1547-4399-5003
BDY83A	BD132
BF108	BF109-110-114-117-156-157-178, BFY43, MM2259, SE7016, SK3045, SFT186, 40349, 2N3114
BF109	BF110-114-117-156-157-178-257, BFY43, MM2259, SE7016, SK3045, 40349, 2N3114
BF110	BF111-114-117-119-140-157-177-178-179A-186-294, MM2259, SE7016, SK3045, 40412
BF111	BD115-127, BF118-178-179B/C-186-257-258, MM3001, SE7001, SK3045, SFB178, 2SC154C, 40412
BF114	BF108-110-115-117-157-178-257, BFY43, MM2259, SE7016, SK3045, 2N3114-5184
BF115	BF117-122-173-184-194-224-225-237, SE1001-5056, SK3019-3122, 40239, 2SC454, 2N915-2952-3693
BF117	BF110-157-178-257, BFY43, MM2259, SE7001-7016, SK3045, 2SC154C, 2N3114-5184
BF118	BD115, BF111-115-178-179C-258-338, MM3003, SE7056, SK3045, 40412, 2SC154C, SBF258
BF119	BF111-178-257-337, MM2269, SE7017, SK3045, 40346, 2N3114
BF120	BF179-258, MJ420, SE7055, 40412, 2N5838
BF121	BF159-167-196-198-224-225-596, BFX60, GI3792, SE5056, SK3039, 2SC454, 2N3137-3932.5236
BF123	BF173-176-197-198-224/JF-237, GI3693, MM8000, SK3039, 2SC464, 2N3689-3932
BF125	BF158-173-197-198-224/JF-237-314, GI3793, SE5031, SK3039, 2N3137-3932-5236
BF127	BF167-176-196-197-198-224-225/JF, BFX60, GI3792, MM8000, SE5056, SK3039, 2SC682, 2N3689-3932

BF130	BFY39
BF131	BF155-181-182-224-311, BFX62, BFY37-39-88, MPS6542, SE3005, 4023Z, 2N3692
BF132	BCY58, BFY37
BF133	BF162-183-240, BFX60, BFY37-39-88, BSX48, ME5001, MPS6541, SE5001, TIS87, 4023Z, 2N3688
BF134	BF115-160-173-200-224-385, BFX59-62, BFY66, ME3001, MPS6542, SE3001, 2N3600-3691
BF136	BF167-185-195-225, BFX60
BF137	BF115-173-194-224
BF138	BF136-167-185-195-225, BFX60
BF140	BF114-117-178-257-258, BFY43-57, MM3001, SE7016, SK3045, 40346, 2N257
BF140D	BF118-178-179B-257-258, BFY57, MM3002, SE7017, 40346, 2SC154C
BF152	BF183-200-385, BFX59-62, BFY66, ME3001, MPS6542, SE3001, NKT16229, SK3039, SX18 2N918-3600-3691
BF153	BF121-184-194-185-235-238-254-255, BFY19, ME3002, NKT16229, SE3002, TIS63, SK3039, 2N3692-5180-5224
BF154	BC108A-172A-183A-208A-238A, BF185-196-200-224-255-311, BFX59, MPS6511-6565, NKT12329, SE2001, SK3039, 40238, 2N918-3564-3691
BF155	BF180-225-240, BFX33-55-62, ME5001, MM1803, SE1001, SK3117, 2N3692-3866-4134-5108
BF156	BF110-111-114-117-140-178-257, BFY43, MM2258, SE7002, SK3045, 2N3114-5184
BF157	BD115, BF111-119-170-173-178-179/A/B-257-258, MM2259, SBF257, SE7001, SK3045, 40346, 2SC154C, 2N3114
BF158	BF115-173-199-200-237-385, BFX59, ME3001, MPS918, SE3001, NKT16229, SX18, 40478, 2N3563-3600
BF159	BF173-199-223-232-237-311, BFX60, GI3793, SK3039, SX18, 2N3014-3137-3932-5236
BF160	BF185-195-200-173-225J-237-255-385, BFX59, BFY66, ME3001, MPS6542, SE3001, SK3039, 40478, 2N3600-3691
BF161	BF115-155-167-181-240-255, BFX55-62, BFY88, ME5001, MPS6541, TIS87, 40237, SE5001, SK3117, 2SC707, 2N4134
BF162	BF167-173-200-224-225/J-240, BFX60-62, BFY88, ME5001, MPS6544, SK3039, 40237, 2N3688
BF163	BF196-198-167-225/J-241, BFX60, BFY88, ME5001, MPS6544, SE6002, SK3039, 40237, 2N3689
BF164	BF167-198-224-225/J-241, BFX60, BFY88, ME5001, MPS6544, SE5003, SK3030, 40239, 2N3690
BF165	BF184-185-194-195-234-235-241. BC107A-180A-182A-207A, BFY19, NKT16229, SE1010, SK3039, TIS62, MPS6565, 2N3563/4-5180-5224

BF166	BF115-167-173-200-225-237, BFX62, MM8000, NKT16229, SK3117, 2N3339-3692-3932-4134
BF167	BF189-196-198-225-237, MM8000, NKT16229, SK3117, SE5056, 2SC682, 2N3339-3693-3932
BF168	BF167-173-225, MM8000, SK3117, 2N3689-3693-3932
BF169	BC107B-108/A-182B-183A-208A, BF115, BFY37-72, SE5006, SK3117, MPS6520, 2N3866
BF169A	BF115
BF169R	BC107B-115-147B-167B-171B-182B, MPS6520
BF170	BF110-117-118-173-174-179A-186-257-258-178, MM2259, SE7001, 40346, 2N3114
BF173	BF117-123-140-166-197-199-224-225-306, FT129, SE5031, SK3018, 2N3399-3693-3932-4072, 2SC464
BF174	BF110-111-114-119-170-178-179A/B-257, MM2259, SE7001, SK3045, SBF257, 40349 2SC154C, 2N3114
BF175	BF127-167-196-198-225-240, MPS6541, SE5001, SK3117, TIS87, 40237, 2N3688-4012-4134
BF176	BF123-173-223-224/5-232-306-311, MM8000-SK3039, SE5025, 2N3339-3693-3932-3689
BF177	BF108-117-156-178-179A-257, BFY43, MM2258, SK3045, SE7002, 2SC856, 2N3114-5184
BF178	BF108-119-140D-156/7-179A-257, SBF178, SE7001, MM2259, SK3045, 40346, 2SC856, 2N3114
BF179	BF118-148-179A-258, MM3001, SE7055, SK3045, SBF258, 40412
BF179A	BF118-119-179/B/C-257-294, MM3002, SE7017, SK3045, 40412, 2SC154C
BF179B/C	BF118-120-179C-258, MM3003, SE7055, SK3045, 2SC154C, 2N3440
BF180	BF155-224-311, BFY19-88, BFX62, MPS6548, MT1061, SK3039, 2N2616-3600-3691
BF181	BF155-224-311, BFX62, BFY19-88, MPS6543, SE3005, SK3117, 2N3692-5180.
BF182	BF155-224-311, BFX62, BFY37-88, MPS6542, SE3005, SK3117, 40237, 2N3691
BF183	BF155-182-224-311, BFX62, BFY37-88, MPS6543, SE3005, SK3117, 40238, 2N3691
BF184	BF115-121-167-194-234-238-240-254-287-234-384, GI3694, MPS3694, SK3117, 2SC460, 2N3932-4437
BF185	BF115-125-167-195-235-237-241-255-288-385, GI3693, MPS3693, SE5056, SK3117, 40478, 2SC535, 2N3932-4436
BF186	BF111-118-120-174-178-179A-225-258, MM3001, SE7055, SK3045, SBF258, 40412, 2SC154C
BF187	BF115-167-173-184-224/5, GI3693, MPS3693, SK3122, 2N3692-3932
BF194	BF115-121-163-173-184-189-197-224-234-238-247B-594, GI3694, MPS3694, SK3018, 40238, SE5056, 2SC460, 2N3694

BF195	BF115-125-163-185-196-173-225-235-237-273/D-595, GI3693 MPS3693, SK3018, SE5056, 2SC535, 2N3693-3932
BF196	BF123-167-198-224/5-270-329-596, MPS6547, SE5002, SK3039, 2SC682, 2N3691-3932
BF197	BF127-173-199-224/5-271-330-597, MPS6546, SE5006-5030B, SK3039, 2SC464, 2N3632-3692
BF198	BF125-163-167-200-224/5-270, MPS6546, SE5056, SK3039, 2N3692-3933-5236
BF199	BF127-173/4-197-200-224/5-271, MPS6546, SE5006, 5030B, 2N3693-3933
BF200	BF162-173-225-271-311-314, SE5020/1/2/3/4, SK3018, 2N2616-2708-3288-3693
BF214	BF115-123-184-194-234-238-251, MPS3694, SE1002, SK3117, 2N3694-5108
BF215	BF123-115-185-195-235-237; BFW63, MPS6565, SE1001, SK3117, 2N3692-5108
BF222	BF115-185-195-235-237, BFX62, SBF222, SE5006, SK3117, TIS87, 2N2707A/B-3693
BF223	BF115=123-197-224-232-234-169-311-385, BFX21-60, MPS6511, SE5030B, SK3039, 40235, 2N915-3014
BF224	BF123-173-199-232-287, MPS6545, SE5030B, SK3039, 40238, 2SC682, 2N916-3339
BF225	BF115-127-163-167-196-198-224-235-288-335, MPS6545, SE5056, SK3039, 40478, 2SC682, 2N918-3339-3866
BF226	BF127-167-185-195-226-235-237-287, SK3117, 2N3301-3693-3933-5236
BF227	BF123-137-173-199-224/5-271, BFR27, 2N3337-3693-3933-5830
BF228	BF177, BFY80, BSW69, BSX21, 2N5830
BF229	BF163-184-194-238-254, SE5056, SK3039
BF230	BF163-185-195-237-255, SE5056, SK3039
BF232	BF173-223-224, SK3117, 2SC464
BF233	BF121-184-234-237-254-274B, 40238, 2N3694
BF234	BF125-184-195-229-237/8-254-274B, SK3039, 40238, 2N3694
BF235	BF121-184/5-195-237-255-274C, SK3039, 2N3564-3693-3933
BF236	BF115-123-235-241-255-287, MPS6565, SE1001, 40238, 2N3691
BF237	BF115-121-185-194-195-163-235-241-287, MPS6565, SE1001, SK3039, 40238, 2N3691
BF238	BF115-121-163-184-194-195-234-240-288, MPS6566, SE1002-5056, SK3039, 40239, 2N3690
BF240	BF115-121-167-196-198-225-234-238-288, MPS6566, SE1002-5056, SK3039, 40238, 2N3691
BF241	BF115-123-173-196-198-224-235-237-287, MPS6565, SE1001-5030B, SK3030, 40200, 2N3091
BF243	AF106-201, 2N3588
BF244	BF245, E304, SK3112
BF244A	E305
BF244B/C	E304
BF245	BF244, E304, SK3112

BF245A	E305, SK3116
BF245B/C	E304
BF246	BF247, E212, SK3112
BF246A	E113
BF246B	E112
BF246C	E111
BF247	BF246, E212, SK3112
BF247A	E113
BF247B	E112
BF247C	E111
BF251	BF127-167-196-225-232, MPS6544, NKT16229, SE5001-5056, SK3117, 2N3693-3932
BF254	BF125-163-184-194-225-234-238-240-273C, MPS6566, NKT16229, SE2002-5056, SK3039, 2N3693-3932
BF255	BF121-163-185-195-224-235-237-241-273D, MPS6565, NKT12329, SE2001-5056, SK3039, 40238, 2N3691
BF256A	BFR80, E305
BF256B	BFR80, E304
BF256C	E304
BF257	BD115, BF111-115-118-119-157-186-179B, MM2259, SBF257, SE7001, SK3045, 40346, 2N3114
BF258	BD115-118, BF118-179C-259-338, MM3003, SBF258, SE7056, SK3045, 2N3440
BF259	BF258-338, MJ421, SE7056, SK3045, 2N3440
BF260	BF162-167-173-200-224-240, BFX60, BFY39-88, ME5001, MPS6543, SE5006, SK3117, 2N3600-3691
BF261	BF167-173-127-196-225-287, MPS6544, SE5003-5056, SK3117, 40237, 2N3690
BF268	BFY90
BF270	BF127-167-196-225, MPS6544, SE5003-5056, SK3117, 40239, 2N3688
BF271	BF123-171-173-224-311, FT129, SE5030B, SK3117, 2N2708-3399-3932-4072
BF272	MM4048, NKT20339, SK3118, 2N2906-3251-5138-5365
BF273	BF224JA, SK3117
BF273C	BF123-184-234-238-241-254, MPS3694, NKT13329, SE1002, SK3117, 2N3694-3932
BF273D	BF127-185-235-237-240-255 ' ¯PS3693, NKT16229, SE1001, SK3117, 2N369' /932
BF274	BF224JA, SK3117
BF274B	BF123-184-234-236-241-2F MPS3694, NKT16229, SE1002, SK3117, 2N3′ 3-3932
BF274C	BF127-185-235-237-240-255, MPS3693, NKT13329, SE1001, SK3117, 2N3694-3933
BF287	BF115-121-167-235-237-240, MPS5657D, SE1001, SK3117, 2N916-4876
BF288	BF115-125-163-167-234-241-236. MPS6539, NKT13329, SE1001-5056, SK3117, 40235, 2N4876

BF290	SE5031, SK3117
BF291	2N915
BF292A	SE7001, SK3045
BF292B	SBF178, SK3045
BF292C	SE7055, SK3045
BF294	BD115, BF120-179B-186-257, MM2259, SE7017, SK3045, 40349, 2N3114
BF297	BF111-118-120-179B-186-257-294, MM2260, SE7017, 40349, 2N3114
BF298	BF118-179C-258, MM3003, SE7055, 40412
BF299	BF259, MJ421, SE7056, 2N3440
BF302	BF127-160-167-185-195-196-237-270, FT129, MPS6546, SE5002, SK3117, 40237, 2N3689
BF303	BF123-173-184-194-197-199-238-271, FT129, MPS6539, SE5001, SK3117, 40238, 2N3688
BF304	BF123-173-185-195/6/7-232-237/8-240-288, FT129, MPS6539, SE5001-5031, SK3117, 40238, 2N3688
BF305	BF111-119-179B-186-257/8-294-297-337, MJ420, SE7055, SBF257, SK3045, 2N5185
BF306	BF121-173-196-224-232-237-240-287, FT129, MPS6574. SE1001-5031, SK3117, 40235, 2N3691
BF310	BF115-121-123-181-234-237-241-287, MPS6574, SE1001-5031, SK3039, 40237, 2SC856, 2N3688
BF311	BF173-183-224-232-288, BFY19, MPS6542, SE5003-5030B, SK3039, 40237, 2N3689
BF314	BF123-199-200-224-232-287, MPS6542, SE1002, 40238, 2N3688
BF329	BF196-596, SE5056
BF330	BF197-597, SE5031
BF332	BF152-594, SK3039
BF333	BF160-595, SK3039
BF334	BF167-197-222-237-240, MPS6540, SK3039, 2N3793-3932-5236
BF335	BF173-196-222-238-241, MPS6540, SE5056, SK3039, 2N3794-3933-5236
BF340	BFX35-88, MM4048, NKT20329, SK3118, 40319, 2N2906-4916-5365
BF341	AF121, BC360VI, BFX48-88, BSV15VI-45VI, MPS6516, GI3702, SE8540, SK3118, 2N2906-4359-5138-5366
BF342	BC360X, BF450, BFX48-88, BSV15X, GI3702, MPS6517, SE8540, SK3118, 2N2906-4359-5138-5366
BF343	BC360XVI, BF451, BFX48-88, BSV15XVI, GI3702, MPS6518, SE8540, 2N2906-4359-5138-5366
BF348	E211
BF357	BF164, BFW92, BFX60, BFY90, BSY95A, ME3001, MPS6544, SE5001, NKT12329, 40456, 2N915-3693
BF384	BF152-180-183-311, BFX62, BFY37-88, MPS6542, NKT16229, SE3001, 2N3600-3691
BF385	BF153-180/1-311, BFX62, BFY37-88, MPS6542, NKT16229, SE3002, 2N3600-3692

123

BF390	SE7056
BF397	2N3963
BF456/7/8	BD115
BF459	BF338
BFR16/17	BC174B, BCY65E, BSX52A-54A, FT107C, MPS6566, 2N930A-2483/A, TIS97
BFR18/19	BFY55, BSX45VI, BSY53-83, FT3722, ME6101, 2N736-3722-3725-3110-5321
BFR20	BSX45XVI, BFS62, BSY78, ME8003, 2N2219A-2789-3109-5321
BFR21	BSV64, BSW10, BSX46VI, BSY85, ME8002, 2N1890-1893-2193A-2243A-2405-3020
BFR22	2N2102
BFR23	2N4036
BFR24	2N4037
BFR25	BSY68, 2N4390
BFR57	BF336
BFR58	BF337
BFR59	BF338
BFS10	BFS23, BFX33-55, 2N3583-3866-5108
BFS12	BFS94, BSV15VI, BSW75, 2N2907-3444-3467-3762-4037
BFS17	BF180, BFS186, BFW77, BFX59, ME3011, SE3005, SK3039, 40294, 2N915-3571-3929
BFS18	BF152-181-384, BFS62, BFX60, BFY19, ME3001, MPS6542, SE3001, SK3039, 2N2708-3691
BFS19	BF153-183-385, BFS62, BFX60, BFY19, MPS6542, SE3002, SK3039, 2N2708-3692
BFS20	BF184-181-224, BFS62, BFX60, BSY95A, ME3001, MPS6541, SE4001, SK3039, 2N2708-3693
BFS21/A	E203
BFS22	40280, 2N3948
BFS23	40290, 2N3948
BFS36	2N930
BFS36A	2N929
BFS37	2N2605
BFS37A	2N2604
BFS38	ZT82
BFS38A	ZT80
BFS39	ZT83
BFS40	ZT182
BFS40A	ZT180
BFS41	ZT183
BFS42	2N2221
BFS43	2N2222A
BFS44	2N2906
BFS45	2N2907A
BFS46	2N918
BFS47	BF111-177, BSW65, BSX21, BSY55, 2N699-1890-2243A
BFS48	BFS93, BSV15VI, ME0401, 2N2243A-2405-2905A-3996-3565-4031
BFS50	BFS10-22, BFX55, ME5001, 40280, 2N3866-3924

BFS51	BFS10-23, BFX59, BSY56-82, ME6101, 40405, 2N3506-3566-3866-5025
BFS52	BFX87, BSV15VI, BSW74, ME0401, 2N2696-2904-2906-3638
BFS53	BFX87, BSV16VI, BSW74, ME0402, 2N2905-2907-2927-3628A
BFS59	BFR51
BFS60	· BFR41
BFS61	BFR39
BFS62	BF225, BFS20, BFX59, BFY79, BSX51, BSW82, ME9001, MPS6542, SE3001, 2N3600-3692
BFS65	BFS69
BFS68/P	E304
BFS86	BF180, BFS10-17, BFX59, ME3011, 40408, 2N915-1722-2949
BFS89	BD115, BF259, 2N3440-3742, SE7056
BFS90	· BF258, 2N3634-4929-5323
BFS91	BF257, BFS93, BSV16VI, 2N2906A-3494-4031-5323
BFS92	BSV16VI, 2N2907A-4031-4147-5149-5322
BFS93	BSV16X, 2N2907A-4031-4149-5123-5322
BFS94	BSV16X, 2N2907A-3634-4040, 40406
BFS95	BSV15VI, 2N2907A-2800-2837-4030-4890
BFS96	BFR61
BFS97	BFR81
BFS98	BFR79
BFS99	2N5830
BFW10	E304, 2N4416
BFW11	E305, 2N4416
BFW12	E305
BFW13	E201
BFW16	· BFS50, BFW19, BFY99, 2SC890, MT1038, 2N3553-3866-4428-4875-5108-5236
BFW16A	BFS50, BFW19, BFY39-40-99, 2SC890, 2N918-3553-3866-4428-4875-5108-5236
BFW17	BFS90, BFW19, BFY99, MT1038, 2SC890, 2N3553-3866-4876-5236
BFW17A	BFS50, BFW19, BFY39-99, 2SC890, 2N918-3553-3866-4876-5236
BFW19	· BFS50, BFY99, 2SC890, 2N3137-3553-3866
BFW20	BC266B, BFX29, BSV16VI, BSW22A, ME0461, 2N3644-3798-3962
BFW21	2N3798/9-3963-4037
BFW22	BCY79C, 2N3505-3644-3964-5087-5367
BFW23	BC266B, BFX30, BSV16XVI, BSW22A, ME0461, 2N3799-3965
BFW24	BC141VI, BFX84, BSY55, ME8002, 2N1893-2243-3020-3108
BFW25	· BC141XVI, BFY55/6, BSX45XVI, ME8003, 2N1613-3109
BFW26	BC141VI-211, BFY55, BSX45VI, ME8002, 2N1711-3110
BFW29	BFY46-68-72, BSY54-71, ME6101, 2N1986-2219A-· 3736-5145

BFW30	BFW77. BFX59, BFY37-69-90, ME1001, MT1061A, EN744 , MPS6507, 2N918-3544-3571-3600-5761
BFW31	BC161VI-327, GI3703, MPS3703, 40406, 2N2904-2907-3703
BFW32	BC141VI-337, GI3705, MPS3705, 4 07, 2N2218-2222-3705
BFW33	BC141VI, BSX45VI, BSY56, ME8002, 2N739-1893-2243/A-2405-3565
BFW36	BF118-178-186-257-258, MM2260, 2N2008-3114, 40340
BFW37	BF111-177-257, MM2258, SE7056, SK3045, TIS101, 40349, 2N3114
BFW39	2N2915
BFW39A	2N2919A
BFW40A	2N2916
BFW41	2N918
BFW43	BF398, MM4001, 2N3494-3930-4888
BFW44	BF398, MM4001, 2N3495-3931-4929
BFW45	BF118-177-257, MM1812, SE7001, SK3045, 40346, 2N2008-3114
BFW46	BSX72, 40280, 25C890, 2N3553-3924
BFW47	BFS86, BFY99, 40478, 25C890, 2N719A-3553
BFW51	2N2974
BFW51A	2N2978
BFW52	2N2975
BFW52A	2N2979
BFW54/5/6	E304
BFW57	BC141XVI, BFR40, BFW26, BFY55, BSX45XVI, ME8003, SE6020, 2N6020, 2N736-1711-2219A-3110-3569
BFW58	BC141VI, BFR40, BFW25, BFY56, BSX45VI, ME8002, SE6020, 2N735-1613-3109-3568
BFW59	BC340XVI-337, GI3705, ME6002, MPS3705, SE6020, 2N2219A-3704/5-4954-5450
BFW60	BC147A-337-340XVI, BFW59, ME6001, MPS3704, GI3704, SE6020, 2N3704/5-5449
BFW61	E304, 2N4222
BFW63	BF167-198-225J-235, FT129, MPS6546, SE5002, 2N3691-3932
BFW64	BF173-224-225J, MPS6539, MT1061, SE5001, 2N3692-3933
BFW65	BF225J
BFW66	BC340XVI, BCY65E, BSX52A, MPS6566, 2N2219A-2483/4
BFW67	2N5058
BFW68	BFW47, BFX55, 40290, 2N2221-2222A-3553-3866
BFW69	BFY99, 25C890, 2N3866-4428-4875-5108-5236
BFW70	BF125-198-255-311, BFX60, MPS6511, 40235, 2N915-3014
BFW71	BFS50, BFW16, BFY99, ME8101, 2SC890, 2N2222A-3866-4428-4875-5108-5236
BFW73/A	BFS50, BFW16A, BFY99, ME8101, 2SC890, 2N3866-4876,5106,5108

BFW74	BFS50, BFW17A, BFY99, ME8101, 2SC890, 2N3866-4876-5106-5108
FW75/6/6A	BFS86, BFW17, BFY99, ME8101, 2SC890, 2N3866-4428-4875-5106-5108
FW77/7A/8	BFS50, BFW16, BFY99, ME8101, 2SC890, 2N3866-4428-4876-5106-5108
BFW87/8	BFR80
BFW89/90	BCW37
BFW91	2N3702
BFW96	2N4416
BFW97	ZXT320
BFW99S	BFW99
BFX11	2N3726
BFX12	BC178/VI-205-213-262, BSW19-21-72, BSX36, MPS6518, NKT13429, 2SA548, 2N869A-2411-2894-5187
BFX13	BC178/VI-213-262, BFX48, BSW19-22-73, MPS6518, NKT13429, 2SA548, 2N869A-2412-2794-5187
BFX15	2N2060
BFX17	BFY70, BSX49-73/4, 40347, SK3024, 2N918-2194-2218A-2270-3725
BFX18	BF167-173-224/5-311-314, BFX60, GI3708, NKT16229, SK3039, TIS85, 2N918-2708-3588-4134
BFX19	BF109-173-234-254-311-314, BFX60, GI3708, NKT16229, SK3039, TIS84-86, 2N918-2708-3690-4134/5
BFX20	BF173-183-311, BFX33-60-62, BFY37, GI3708, MPS6543, NKT16229, SK3039, TIS84-86, 2N918-2708-4135
BFX21	BF173-181-311, BFX32-59-62, GI3708, NKT16229, SK3039, TIS84-86, 2N918-2708-4135
BFX25	2N3504
BFX29	BC161-361VI, BFY74A, BSV16VI, ME0401, 40406, 2SA537, 2N2904A/5A-3072-3485A-3645-4030
BFX30	BC161-361XVI, BFX41, BFS91, BSV16VI, ME0401, 40406, 2SA537, 2N2905A-2907-3645-4032-4356, 4404
BFX31	BC173, BF180-240, BFX21-60-62, BSX53, MPS6543, GI3707, NKT16229, SK3117, TIS86, 2N2708-4134/5
BFX32	BF123-180-311, BFX59, MPS6546, SK3039, 2N2708-3251-3866-3932-4135-4874
BFX33	BFX55, BFW17, 2N2218-3137-3866-5108
BFX34	BSV64, BSX46VI-85, SK3045, 2N2193A-3419-4300-4895-5154
BFX35	BC360XVI, BCY79B, BSW21A, 2N3251-3504-3644
BFX36	2N3810-4024
BFX37	BC161XVI, MM4000, 2N2907A-3962/3-4033-4928-5323
BFX38	BC161/X, SK3025, 2N3468-3763-4032-4037-4356, BFT80

127

BFX39	BC360VI, BSV15VI, BFT80, BSV15VI, BSW224, ME0401, 2N3468-3763-4030-4037-4354
BFX40	BC161/VI-313, BFT79, BSV16VI, BSY40, ME0401, SK3025, 2N4031-4033-4236-4356-5323
BFX41	BC161/XVI-313, BFT79, BSV16XVI, BSW19A, BSY41, SK3025, 2N4030/1-4236-4356-5323
BFX42	BFW30, BFX32-59-73, ME3002, MM1500, 2N915-918-3839-4251
BFX43	BFX92, BFY90, BSX19-92, BSY63-95A, ME9001, SK3039, 2N2368/9-2616-3563-4072-4873-5180
BFX44	BFX93, BFY90, BSX20-93, BSY63-95A, ME9001, NKT16229, SK3039, 2N2368/9-4134-4874, 40405
BFX45	BC148, BCW49, BCY58C, BSW42A, BSX24-38-81-93, MPS6542, NKT12429, SK3039, 2SC907, 2N2222-2369-3691-3932
BFX48	BC179, BCY78A, BSW21, ME0463, NKT20329, 40406, 2N2905-4034-4061-4125-4916-5208
BFX50	BFX61-68A, BFY55-56A, BSY83, SK3024, 2N1889-2193-2222A-2297-3110-4014-5231
BFX51	BFX61-68A, BFY55-56A, BSY83, SK3024, 2N1889-2193-2221A-2297-3108-4047
BFX52	BFX72, BFY52-67C, BSX22-32, SK3024, 2N2195/B-2222A
BFX53	BFY90
BFX55	BFS10-23, BFY88, SE3006, 2N2218A-3553-3866, 40347
BFX59	BFS17, BFX73-89, BFY90, FT129, ME3011, NKT16229, SE3005, 2SC890, 2N918-3794-3839
BFX60	BF173-198/9-271, FT129, NKT16229, SK3117, 2SC463, 2N918-3337-3693-3933-4876
BFX61	BFX51-68A, BFY50-55, SFT443, SK3024, 2N1889-1893-2193-2297-3110-4014-5321
BFX62	BF180-224-270, BFY90, ME3001, MPS6542, NKT10519, SE5025, SK3117, 2SC707, 2N918-2708-3691
BFX65	2N3962
BFX66	2N997/8
BFX67	2N997
BFX68	BFY68, BSX45XVI, BSY54-71, SK3024, 2N1711-2049-3019-3253-3735-3737-5262
BFX68A	BSX45XVI, BSY54, SK3024, 2N1711-2193-3019-3109-3252
BFX69	BFY67, BSX45VI, BSY53, SK3024, 2N1613-3020-3170-3253-3735-5262
BFX69A	BFY55, BSX45VI, BSY54-83, SK3024, 2N2193-3020-3110-3253
BFX70	2N2060
BFX71	2N2223
BFX72	2N2223A
BFX73	BFX58-89, BFY90, ME3011, SE3005, SK3117, 2N918-3600-3839
BFX74	BC161VI, BFX87, SK3024, 2N1131-2904-3133

BFX74A	BC161VI, BFX29, SK3025, 2N1132-2800-2837-2904A-3072
BFX77	BF173, BFY74, BSW61, BSX48-79, MPS6532, SK3117, 2N918-1986-2221-3933-4951-5144-5450
BFX82	2N4381
BFX83	2N4382
BFX84	BFW24, BSX46VI, BSY85, ME8002, SK3024, 2N1889/90-3056-3108-5320
BFX85	BFW33, BSX46VI, BSY86, ME8002, SK3024, 2N1890-2405-3057-3107
BFX86	BFY46-50, BSX32-73, BSY83, SK3024, 2N2195B-2218-3109-3507-3725-5189
BFX87	BC161VI, BFX74A, BSW21A, ME0401, SK3025, 2N2904/A-3468-4037-4890
BFX88	BC160VI, BFX74A, BSW21A, ME0402, SK3025, 2N2904-3467-4037-5042
BFX89	BF357, BFX20-62-73, BFY66, ME3011, MT1061, 2SC707, 2N2616-2857-918-3839, SE3005
BFX90	2N3930-4357-4930
BFX91	2N3931-4358, MM4002
BFX92	BCY59A, BSY77-93, ME4101, SK3122, 2N929
BFX92A	BCY65E, BSY93, ME4102, MM2483, 2N2483
BFX93	BCY59B, BSY77-93, ME4104, SK3122, 2N930
BFX93A	BCY65E, BSY78-93, ME4102, MM2484, 2N2484
BFX94	BFY34, BSW61-84, BSX73, ME6101, SK3024, 2N2221-3301-4951
BFX95	BFY46, BSW62-85, BSX74, ME6102, SK3024, 2N2222-3302-4952
BFX96	BFX97, BFY34, BSX73, BSW51, BSY53, ME6101, SK3024, 2N2218-3299
BFX97	BFY46, BSW52, BSX74, BSY54, ME6102, SK3024, 2N2219-3300
BFX98	BF111-119-178-257, MM3001, SE7017, 40412, 2N3114
BFX99	2N2060/B
BFY10	BC107, BFX45, BFY33-51-76, BSX48-75, BSY51, SK3018, TIS87, 40239, 2N840-929-1565-3693
BFY11	BC107, BFY27-33-51-77, BSX48, ME5001, SK3018, TIS87, 2N841-915-929/30-1566-3693
BFY12	BCY10, BSX73, SK3024, 2N2218A-2219/A, 2SC479
BFY12B	BFX95, BFY50-70, BSX49-73, ME4101, 40317, 2N2194-2217-2218A-2787
BFY12C	BFX55-96, BSW53-84, BSX74, ME4101, 40314, 2N2193-2218/A-2788
BFY12D	BFX97, BFY46, BSW54-85, BSX74, ME4101, 40459, 2N2192-2219/A-2789
BFY13	SK3024, 2N2219/A
BFY13B	BC211, BFX68, BFY44, BSW10, BSY87, ME8003, 40348, 2N736-2210A-3567-3722-4945
BFY13C	BC211, BFX69, BFY44, BSW10, BSY87, ME8003, 40348, 2N736-2218A-3568-3725-4945

BFY13D	BC211, BFX69A, BFY44, BSW10, BSY88, ME8003, 2N736-2219A-3569-4047-4946-5321
BFY14	BF140, BFY43, SK3024
BFY14B	BFR21, BSV64, BSY45-55, ME8002, 40367, 2N698-719A-739-3419-3701
BFY14C	BFR21, BSV64, BSY45-55, ME8002, 40408, 2N699-720A-740-3421-3701
BFY14D	BSV64, BSY45-56, ME8002, 40408, 2N740-3107-3499-3700-4410
BFY15	BFY33-67, BSW19, SK3024, 2N697-930-2219
BFY16	BFY12, SK3024, 2N697-2219
BFY17	BFY12-39-54, BFX19-55, BSX72, BSW51, ME6001, SK3024, 2N918-708-2218/9-2222-3137-3261-3508-4148-4875
BFY18	BCY59, BFX19-75, BFY19-39-51-84, BSW41, BSX39-48-75, BSY93, SK3025, 2N708-918-2222-3014-3261-3508-4148-5451
BFY19	BFX20, BC108A, BSW62, BSX38-48, BSY19-33-63, ME9001, MPS6542, SK3024, 2N708-753-930-2222-3691-3932-4137-4874-5135
BFY20	BF173, 2N2218-2640
BFY22	BC121/W-146R, BF224, BFY47R-87Y, SK3020, 2SC475, 2N930
BFY23	BC121/Y-146R, BF224, BFY47Y-87Y/G, SK3020, 2SC476, 2N930
BFY23A	BC121G-146G/BN, BF224, BFY87W, 2SC476, SK3020
BFY24	BC121/W-146/R/GRT, BFY47R-87A, SK3020, 2N918-930
BFY25	BFY28, BSY34, SK3122, 2N2219
BFY26	BFW66, BSW63-84, BSX49-79, BSY63, SK3122, 2N708-2194-2219A-3253-3642-3903-3946-4140-5189
BFY27	BF167, BFY28-46, BSW63, ME6101, SK3024, 2N718A-915-918-2222/a-2221-2788
BFY28	BSY34, SK3122, 2N2221
BFY29	BC123-146R, BFY49R, SK3020, 2N930
BFY30	BC130-146R, BFY49R, SK3020, 2N930
BFY33	BFX17-33, BFY67, BSX46-95, BSY44-53-92, ME6101, MPS3705, SK3122, 40451, 2SC479, 2N697-1613-1984-3705-4046-5450
BFY34	BFR19, BFY68, BSX45-95, BSY11-44-54, ME6102, SK3024, 2SC479, 2N1613-2218A-2787
BFY37	BC107-108A, BFX31, BSX38-48-51, BSY70-93-95A, NKT12329, SK3019, 40454, 2SC907, 2N706A-835-2218-2222, ME9001
BFY39	BC107-182, BCY58, BFY27, BSY39, MPS6565, SK3020, 2N930
BFY39I	BC107A-182, BFY39, BSW54A, BSX33-48-51A-54A, NKT10439, SK3122, 40398, 2SC284, 2N834-916-4148
BFY39II	BC107A/B, BFY39, BSW41, BSX33-39-45X-51A-54A, NKT10339, SE6001, 40398, 2N3705-4123-4876

BFY39III	BC107B/C, BFY39, BSW41, BSX45XVI-52A-54B, MPS6532, SE6001, NKT10439, 40397, 2N3704-4876
BFY40	BC140/VI, BFX96, BFY34-67, BF178, BSW51, BSX45-73, BSY44-92, ME6102, SK3024, 2SC53-708, 2N1313-1613-2410-3053-3299-5189
BFY41	BC141/VI, BF111-140-257/8, BSW67, BSX46, BSY45-55, ME8002, MM2258, 40349, 2N698-1889-1893-2218A-2443-3498-4001, 2SC708A
BFY42	BC107
BFY43	BF110-114-117-140-178-257/8, BFX98, BFY45, BSW68, BSY56, ME8002, SK3045, 40346, 2SC154C, 2N3114-3500-3712-3923
BFY44	BFW59, BFX51, BFY12-50, BSW10, ME8003, SFT443, SK3024, 40348, 2N2193-3444-3507-3553-3665-3725
BFY45	BF178, BFS90, BFW36, BFY43-57-65, BSW66-68, BSY56, ME8002, SE7002-7016, SK3045, 40346, 2SC857, 2N1893-1990-3114-3500
BFY46	BFY68, BSX45-96, BSY54-71, ME6102, SK3024, 2SC479, 2N1711-2219A-2789
BFY47	BC121, SK3018
BFY48	BC122, SK3018, 2N930
BFY49	SK3018, 2N930
BFY50	BFR18, BFY68A, BSX45VI-46, BSY46-85, ME6001, SK3024, 40348, 2SC708, 2N2193-2222-2297-3110-3253-3444-4047
BFY51	BFX51-68A, BFR19, BSX45/X, BSY46-85, SK3024, 40347, 2SC708, ME6102, 2N2193/4-2218-2410-3053-3252-4046
BFY52	BFX52-72, BFY56, BSX33-45/XVI, BSY46, ME8001, SK3024, 2SC708, 2N2195-2219-3053-3252-3724-4427
BFY53	BFX52-72, BFY46-50-52, BSX46, ME8001, SK3024, MM3724, 2N2195-3724A-3924-5023
BFY55	BFR18, BFX51-68A, BFY50, BSX45/VI, BSY46-83, ME6001, SK3024, 40348, 2SC708, 2N2193-2297-3110-3444
BFY56	BFR18, BFX50-68A, BFY50, BSX46-61, BSY83, ME6001, SK3024, 40348, 2N2193-2297-3110-3252-3444
BFY56A	BFR18, BFX50-68A, BFY56A, M6001, SK3024, 40348, BSW65, 2N2193-2195-2297-3108-3110-3444
BFY57	BF110-117-140, BFS50, BFX98, BFY41, BSW67, BSY55, ME8002, SK3045, 40346, SE7002, 2N3114-3500-3923
BFY63	BFX62-89, BFY90, ME3011, SE3005, SK3122, 2N918-2616-3137-3478-3570-3600-3839
BFY64	BC327-360XVI, BSV15VI, BSW21A, ME0402, 40406, 2N3120-3244-3467-3644-5042
BFY65	BC141, BF177, BFR21, BFY14-41-43-45-80, BSX46, BSY55/6, ME8002, SK3045, 2SC154, 2N699B-1893-1990/N-2243-3036-5320

BFY66	BF180, BFX62-73-89, BFY56-63, ME3011, MPS918, SE3005, SK3024, 2N918-2616-3563
BFY67	BFR19, BFY34-57, BSX45/X, BSY44-52-53, ME6101, SE8002, SK3024, 2SC408A, 2N915-1613-3735
BFY67A	BFR19, BFY34, BSX45VI, BSY44-53, ME6101, SE6020, SK3024, 40347, 2SC708, 2N2219-2270-3053
BFY67C	BFX17, BSX45VI-49-74, BSY53, ME6101, SE8001, SK3024, 2SC708, 2N2219-3724-3866-4013
BFY68	BFR20, BFY46, BSX45/XVI, BSY54-71, ME6102, SK3024, 2N1711-2049-5321
BFY 68A	BFR20, BFY46, BSX45X, BSY54-71, ME6102, SK3024, SE6002, 2N2219-3947
BFY 69/A	BC122-146, SK3039, 2N706A
BFY 70	BC140, BFX95, BFY50, BSW10, BSX73, ME8003, SFT443, SK3024, 2N1613-2193/4-2217-2787-3724
BFY 72	BFY33-67/C, BSY44-53, ME6101, SK3024, 2N2218/A-2219-3724-3866-4013, SE8001
BFY73	SK3122, 2N2219
BFY74	BFW47, BFX55, BFY27, ME5001, MPS3826, SK3122, 2N915-2221-2483-2586-4994
BFY75	BC107A, BFW47, BFX55, BFY27, ME5001, MPS3826, SK3122, 2N915-2221-2222A-2483-2586-4994
BFY76	BCY59B-66, BSX51A, SE1001, SK3122, 40235, 2N929-2483-3693
BFY77	BCY59IX/C-66, BSX52A, SE1001, SK3122, 40237, 2N929/30-2484-3494
BFY78	BC109, BFW30, BFX55, BFY90, ME1001, MPS6579, SE3005, SK3039, 2N918-3563-4997-5179-5761
BFY79	BFY19, BSX48, ME9001, SE5001, SK3117, TIS84, 40482, 2N2708-3737-4134-4137-4874
BFY80	BFR21, BF177, BFY14-41-43, BSX21-46-46VI, BSY55/6, ME8002, SK3045, 2SC857, 2N719A-1893-699B-1490-1990/N-2643-2896-3036-5320
BFY81	2N2917
BFY82	2N2982
BFY83	2N2223
BFY84	2N3423
BFY85	BCY87, BF184, BFY19, 2SC280, 2N2640-2896-3036
BFY86	BCY89, BF185, BFY19, 2N2639-3036
BFY87	BC122-146, BF185, BFY19, 2N930-5181
BFY87A	BC122-146
BFY88	BF185, BFW17, BFX20-60-89, BFY19-90, BSX88A, FT129, SE5030B, 2SC707-890, 2N918-2616-3553-3600-3691-3866-3933-4876-5181
BFY90	BF169, BFX20-59-63, BFY63, ME1001, MPS6579, MT1061, SE3005, SK3039, 2N918-2222-3015-3563-3571-3600-4875-5180-5761
BFY91	BCY59VII, BF169, BFY86, 2SC280, 2N2642-2915
BFY92	BCY59VII, BF169, BFY85, 2SC280, 2N2643

BFY93	2N2920
BFY94	2N2904
BFY95	2N2906
BFY99	BFS50, BFW16-69, SFT443, SK3024, 2N706A-3053-3553-3642
BFZ10	BCZ11
BLX82	2N5039
BLX83/4	2N5038
BLY12	BD111, BDY12-16A-34-62, SK3027, 40347, 2N3297-3713-4300-4914
BLY14	BLY15A-22-28-74, 2N3375-4040-3553, SFT443
BLY15	SFT443A, SK3027
BLY15A	BLY22-27-74, 2N2632
BLY16	SFT443A
BLY17	BLY16-72, SFT441A, 3TE120, 2N3772-4301-5008
BLY20	BDY15A, BLY16-74-81, SFT443A, 2N1208-3375-3632-4040-5083
BLY21	BDY16A, BLY16-74, SFT443A, 2N1212-3632-4040-5084
BLY22	BD137, BDY16B, BLY74-81, SFT440, 2N1208-3375-3632-4040-5085
BLY25	2N4116
BLY26	2N4115
BLY28	BD139, BLY16-33-74, SFT443A, 2N3632-4041
BLY29	2N4075
BLY30	2N4076
BLY33	BLY22-34, MM8003, MSA8508, 2SC598, 2N3553-3866
BLY34	BDY34, BLY35-74, MPSU02, PPR1008, MS8508, 3TE130, 2SC637, 2N3866
BLY36	BLY22-37-76, MSA8506, PPR1006, 3TE220, 2SC637, 2N3632-4012
BLY37	BLY38-61-76, MM8003, 2SC598, 2N1485-3632-3866
BLY38	BLY40-72, 2N1486-3375-4003-5008-5429
BLY40	BD117, BLY47, 3TE120, 2N1724A-3772-5002-5429
BLY47	BD117, BDY55, BLY47A, 3TE120, 2N3772-4915-5004-5429
BLY48A	BDY55, BLY49, BUY18, 3TE130, 2N2822-3080-3772-5264
BLY49/A	BUY18, 2N2822-3080-5264
BLY50	BLY50A, BUY18, 2N2822-3080-5264
BLY50A	BLY53-74, 2N2822-3080-3375-3733-4012
BLY53	BLY57-62-74, 2SC637, 2N1483-3375-4012-5083
BLY55	MSA8507
BLY57	BLY22-58-63-74, MPSU02, 2SC637, 2N1208-3375-3632-3926
BLY58	BUY22-59-74, 2SC636, 2N1208-3632-3927-4012
BLY59	BLY60-74, 2SC598, 2N1479-3375-3632-3927
BLY60	BLY61-74, 2SC598, 2N1480-3632-3866-3927
BLY61	BLY74, MSA8508, SK3024, 2SC598, 2N1485-3375-3866
BLY62	BLY74, MSA8507, 2SC598, 2N1485-3375-3632-3733
BLY63	BLY74, MSA8506, 2SC598, 2N1485-3378

BLY64	BDY12VI-34-53, 3TE120, 2N3772-3996-5002-5427
BLY65	2N5003-5386
BLY66	BDY13VI-53, BLY68, 3TE120, 2N3863-3996-5004-5083-5428
BLY68	BDY13VI, 3TE130, 2N1724-3864-3997-5008-5069-5429
BLY70	BDY13VI, 3TE130, 2N1724-3772-3998-4915-5085-5427
BLY72	BDY12VI-55, 3TE130, 2N3772-4003-5006-5288
BLY74	BLY22-60-62, 2SC598, 2N3632
BLY75	2N5025
BLY76	2SC598, 2N3632-4012
BLY78	BLY22-59-62-74, 2SC598, 2N3375-3632
BLY79	BLY22-57-63, 2SC636, 2N3632
BLY80	BLY22-63-74-88, 2SC636, 2N3632-3926
BLY81	BLY22-91, 2SC598, 2N3632-3927
BLY87	BLY22-63-74, 2SC636, 2N3375-3926
BLY88	2SC636, 2N3632-3927
BLY89	2SC990, 2N3632-3772-3996-5002-5427
BLY91	2SC636, 2N3632
BLY92	2SC636, 2N3632
BLY93	2SC990, 2N3632-3772-6002
BRY23	FT2011
BRY24	FT2012
BRY25	FT2013
BRY26	FT2014
BRY28	2N1592
BRY30	2N1598
BRY31	2N1599
BRY32	2N4108
BRY33	2N4109
BRY34	2N4110
BRY35	2N4114
BRY36	2N4216
BRY37	2N4218
BSS10	BSX20, 2N3261
BSS11	2N2369A
BSS12	BSX20, 2N3011
BSS13	2N3053-5189
BSS14	BFX34, 2N5262
BSS15	BSV94, 2N5320
BSS16	BSV93, 2N5321
BSS17	2N4036-5322
BSS18	2N4036-5323
BSS19/20	BSW69
BSS30	2N1889
BSS31	2N1890
BSS32	2N1893
BSV15VI	BC360VI, BCY79A, BFY67, BSW19A, 40406, 2N3135-3245-3467-3644-5447
BSV15X	BC360X, BCY79B, BFY67, BSW21A, 40406, 2N3135-3244-3467-3644-5447

BSV15XVI	BC360XVI, BCY79C, BFY67, BSW22A, 40406, 2N3136-3244-3644-5042-5448
BSV16VI	BC327-361VI, BFS12, BSW21A, ME0401, 2N3485-3644-3671-4030
BSV16X	BC327-361X, BFX74A, BSW21A, ME0401, 2N3485/6-3644-3671-4030
BSV16XVI	BC327-361XVI, BFX74A, BSW22A, ME0402, 2N3644-3486-3671-4032
BSV23	ZTX310
BSV24	ZTX311
BSV25	ZTX312
BSV26	ZTX313
BSV27	ZTX314
BSV28	ZTX341
BSV29	ZTX342
BSV33	ZTX510
BSV35	2N2369
BSV35A	2N708
BSV36	2N2475
BSV37	2N2894
BSV38/P	E111
BSV39/P	E112
BSV51	BSX21
BSV52	BC122
BSV62	2N4393
BSV64	BFX34, BSX63VI
BSV82/3	BFT79
BSV89	BSW42-51-82, BSX19, BSY62, GI2926, ME9001, SE1010, 2N2616-3932-4252-4264
BSV90	BSW43-52-83, BSX20, BSY63, GI2926, ME9002, SE3001, 2N3600-3933-4253-4265
BSV91	BSW42-53-48, BSX19, BSY62, GI2926, ME9001, 2N2616-4996-5179-5224
BSV92	BSW43-54-79, BSX79, BSY63, GI2926, ME9002, SE4001, 2N3600-4997-5180-5222
BSW10	BC141VI, BF177, BFX34-84, BSX23-46, ME8002, SK3024, 2N699-1890-2102-2218A-2219A-3107-3722-5320
BSW11	BSW13-59, SE3646, SK3122
BSW12	BSW13, BF173, BSX69, BSY19-63, SK3122, 2N708-914-3646
BSW13	BC146
BSW15	BFT81
BSW16	BFT80
BSW19	BC178, BCY71-78/VII, 40406, 2SA548, 2N2906-3014-3504
BSW19VI	BFX48, BSW21A, BSY40-59, MPS6533, 2N3644-4037-4059/60-5138
BSW19A	BFX48, BSW21A, BSY41-59, MPS6534, 2N3644-4037-4059-4061-5138
BSW20	BC307
BSW20VI	BFX48, BSW44, BSY40-59, MPS6533, 2N3644-4037-4059/60-5138

BSW20A	BFX48, BSW45, BSX41, BSY41-59, MPS6534, 2N3644-4037-4059-4061-5138
BSW21	BC178, BCY72, SK3114, 2N2894-2906
BSW22	BC178, SK3114, 2N2907
BSW22A	BCY79VIII, SK3114
BSW23	BSV16VI, BSX29, ME0401, SK3025, 2N2904-4037
BSW24	BSV16VI, BSX30, ME0401, SK3114, 2N2906-4037
BSW25	BFX12-29, BCY72, BSW19VI-81, ME8201, 2N2894A-3012
BSW26	BFX17, BSW63, BSX45VI-59, ME8003, SK3024, 2N696-2194-2217-2221A-4047
BSW27	BC341VI, BFX17, BSX46VI-59, BSY55, BSW65, SK3024, 2N697-2193-4047
BSW28	BC341VI, BSW65, BSX46VI-59, BSY55, SK3024, 2N731-2194-2218-4047
BSW29	BC340VI, BFX17, BSX45VI-60, BSY53, SK3024, 2N2195-4046
BSW32	BFR21, BFY45-80, BSW70, BSX21, BSY79, ME1120, SK3045, 2N911-1889-1974-1990-5320-5830
BSW33	BC147A-207 , BCW33, BFW64, BFY46, BSW82, BSX22-52A-53, SE6001, SK3122, 2SC641, 2N916-1482-1711-3691-3933-4876
BSW34	BC147A-207, BCW47, BFW68, BFY46, BSW84, BSX23-52A-53, ME5001, SK3122, 2N1482-3693-3866-4014
BSW35	BC147A-207, BCW46, BFX92A, BSW84, BSY44, ME5001, SK3024, 40347, 2N929A-1890-2405-2483-3053-3946-4994
BSW36	2N2905
BSW37	2N2894
BSW38	2N2369
BSW41	BFW64, BSX45VI-48-51A, BSY75-95, ME8001, SE6001, SK3122, 2SC321, 2N916-2221-3693-3866-3933-4876
BSW42	BC207-238A-317, BCY58A, BFY76, BSX24-38A-81, GI3710, MPS3710, SE8040, SK3122, TIS86, 40398, 2N929
BSW42A	BC237A, BCY59A, BFX92A, BSX54A, SE4020, SK3122, TIS87, 40407, 2N929A
BSW43	BC208-238B-318, BCY59C, BFY77, BSX54B, GI3711, MPS3711, SE8040, SK3122, TIS86, 40397, 2N930
BSW43A	BC237B, BCY59C, BFX93A, BSX54B, SE4020, SK3122, TIS87, 40407, 2N930A
BSW44	BC205-308VI/A-321, BCY78A, BFX48-88, BSW20VI, BSX40, MPS6519, SE8540, SK3114, 40406, 2N3250-4059
BSW44A	BCY79A, BFX87, BFY64, BSW20A, BSX40, SK3114, 40406, 2N3250A-3798-4354-5086
BSW45	BC206-308B-322, BCY78C, BFW20, BFX29, BSW20A, BSX41, MPS6522, SE8540, SK3114, 40406, 2N3251-4059
BSW45A	BC327, BCY79C, BFW23, BFX30, BSW20A, BSX41, SK3114, 40406, 2N3251A-3799-4355-5087

BSW50	BFY33-67, BSY44-53-63, 2N2217
BSW51	BFR19, BFY34, BSY34-44-53, 2SC479, 2N1613-2218
BSW52	BFR20, BFY34, BSY34-44-53, 2N1711-2219
BSW53	BFR19, BFY46, BSY34-54-71, 2SC479, 2N1613-2218/A
BSW54	BFR20, BFY46, BSY34-54-71, 2N1711-2219/A
BSW58	BSX72, BSY63, 2N2369
BSW59	BSX72, BSY62, 2N2222-2369
BSW61	BFR18, BFX55, BFY34, BSX49, BSY44-53, 2N1613-2221
BSW62	BFR20, BFX55, BFY34, BSX49, BSY44-53, 2N1711-2222
BSW63/4	BFR18, BFX55, BFY46, DOX49, DOY54-71, 2N1613-2221
BSW65	BFX61, BSW10, BSY47-63VI, SE6021, SK3024, 2N2243A-1715-2405-3019-3725-4000-4236
BSW66	BSX46, SK3024, 2N1716-2405-2988-3019-3498-3712-4001-5338
BSW67	MPSU03, 2N2008-3140-3712-4001
BSW68	2N3400-3500-3712-4929
BSW69	BF119-178, BFW44, BFY45, SE7001, SK3044, TIS101. 40354, 2N3114-3712-4925
BSW70	BC177, BF117-177, BFX98, BSW32, SK3045, 2N870-910-1990-4410-5185
BSW72	BC327XVI, BSV15VI, BSX36, ME0401, SK3114, 40406, 2N721-2906-2837-3135-5447
BSW73	BC327, BSV15X, BSX36, ME0401, SK3114, 40406, 2N722-2838-2907-3136-5448
BSW74	BC327XVI, BFX41, BSV16VI, ME0401, SK3114, 2N2906/A-4027-4031-5322
BSW75	BC327, BFX40, BSV16X, ME0402, SK3114, 2N2907/A-4029-4033-5322
BSW78	BSX88, BSY19-63, ME9001, NKT13329, 2N2368-3210
BSW79	BSX87, BSY19-63, ME9001, NKT13429, 2N2369-3211
BSW80	BSX87A, BSY19-63, ME9001, NKT13429, 2N914-2369-3013-3211-3648
BSW81	BFX12, BSW19VI, BSX29, ME8201, 2N2894A-3546-5055-5292
BSW82	BSW41-61, BSX39-48-72, BSY51, ME6001, MPS3706, NKT10339, SK3122, 2N1986-2221-3261-5451
BSW83	BSW41-61, BSX48/9-72-88A, BSY52, ME6002, MPS3706, NKT10439, SK3122, 2N1987-2222-3261-3706
BSW84	BSW10, BSX33-49-63VI-70, BSY53, ME6101, SK3122, 2N2221/A-2193-2790
BSW85	BFX68, BSW10, BSX49-63X-71, BSY54, ME8003, SK3122, 2N2192/A-2222/A-2791
BSW88	BC167A, BSW33, SK3122, 2N3692
BSW88A	DC108A-183A-208B, BFY39, BSX51A-69, MPS6574, NKT10419, SE8040, 40398, 2N3566
BSW88B	BC168B-183B-208C, BFY39, BSX52A-67, MPS6574, NKT10519, SE8040, 40397, 2N3566

BSW89	BC167A, BSW33, SK3122, 2N3692
BSW89A	BC183A-208B-238A, BFY39, BSX51A-69-81, MPS6574, NKT10419, SE8040, 40398, 2N3566
BSW89B	BC183B-208C-238B, BFY39, BSW88B, BSX52A-81, MPS6574, NKT10519, SE8040, 40397, 2N3566
BSW93	2N5023
BSW96	2N2193A
BSW97	2N2906
BSW98	2N2193
BSW99	2N2904
BSX12	BSX48-72, BSY58-81, SK3039, 2N3303
BSX19	BSX48-75-92, BSY81, ME6001, NKT13329, SK3122, 2SC689, 2N2368-3210-3648-4138
BSX20	BSX48-75-93, BSY19-82, ME6002, NKT13429, SK3122, 2SC689, 2N2369-3211-3648-4138
BSX21	BFS47, BFW23-45, BFY45-80, BSY79, ME1120, SK3045, 2SC857, 2N698-719-1893-1990-5184
BSX22	BC140, BFX96, BFY50, BSW10-26-29, BSX45-60-62VI-72, SK3024, 2N1613-2195-2270-3252-3724-3734-4046-4231
BSX23	BC141, BFR21, BFT39, BFX34, BSW10-26-35, BSX62-63VI, BSY46, SK3024, 2N1480-1613-2896-3019-3036-3056A-3262-3725
BSX24	BC107/A-168A, BCY56-58-59, BFX48, BSW27-41, BSX38/A-51/A-69-81, MPS834, NKT10339, SK3122, 2SC707, 2N834-2221-3692-3933-4876
BSX25	BCY59, BSW41-82, BSX19-22-24-39-48-51A-75, ME6001, SK3122, 2N930-956-2222-2368-3014-4123-4137-4427-4876
BSX26	BSX20-48-75, BSY19-63-82, ME6002, SK3039, 40405, 2N708-914-2368/9-3013-3210
BSX27	BFW30, BFY66, BSX20, BSY17, BSW80, ME3001, SK3039, 2N709-918-2368/9-2475-3010-3493-5186
BSX28	BFY90, BSW80, BSX20, BSY63, ME8101, SE3001, SK3039, NKT16229, 2N918-2368/9-3011-3563-4274
BSX29	BCY78, BFX12, BSW19/VI-81, BSY41, ME8201, TIS53, 2N2894-2906-3012-4258-4313
BSX30	BFY34-67, BSY44-53, ME6102, SK3122, 2N2410-2537-2846-3015-3053-3299-3641
BSX31	2N110
BSX32	BFY34-67, BSW85, BSY34-44-53, ME6102, 2N718-731-2218-2221-2476-2868-3725
BSX33	BFY46-68, BSY54-71, ME8003, SK3024, 2N718A-735-2193-2218A-2221A-3110
BSX35	BFX12, BSW19VI-81, BSX20, ME8201, SK3039, 2N2894A-2906-3304-3639-4207
BSX36	BC360VI, BFX29, BSV15VI, BSW21A-41, BSY59, ME0401, MPS6533, NKT20339, SK3114, 2N2906/7-3644-5042
BSX38	BC107B, BCY58/VI/VII, BSX24-51, BSY19, SK3122, 2SC907, 2N708-2222

BSX38A	BC168A-183A-208A, BFY39, BSW88A, BSX38A-51A-69, MPS6574, NKT10419, SE1001, 40398, 2N3566
BSX38B	BC168B-183B-208B, BFY39, BSW67-88B, BSX52A MPS6574, NKT10519, SE1001, 40397, 2N3566
BSX39	BSX20-48-75-97, BSY82, ME6001, NKT10339, SK3122, 2N708-2710-3014-3705-4013-5450
BSX40	BC139-160, BCY78A, BFX48-88, BSV15VI, BSW21A, BSY59, ME0401, SE8540, SK3025, 40406, 2SA537, 2N2904-3120-3644-4402
BSX41	BC100-139-160, BCY78B, BFX48-88, BSW22A, BSV15XVI, ME0402, SE8540, SK3025, 2N3121 3644-4403
BSX44	BSX12, BSY17-20-38-70, ME8101, 2N709-744-1708-2205-3013-5128-5186
BSX45	BFT41, BFY55, BSX22, BSY46, SK3024, 2N2218A-2297-3252-4231
BSX45VI/X	BC140VI-211, BFR18, BSX95, 40348, 2N2194A-2297-2788-3110-3444-4400
BSX45XVI	BC140XVI, BFR18, BSX96, 40348, 2N2192A-2219A-2789-3109-4401
BSX46	BFT40, BSV64, SK3024, 2N3252
BSX46VI	BC141VI, BFR21, BSY55, ME8002, 40360, 2N870-912-1975-2890-3108
BSX46X	BC141X, BFR21, BSY55, ME8003, 40360, 2N871-911-1974-2891-3108
BSX46XVI	BC141XVI, BFR21, BSY56, 40360, 2N910-1973-3019-3107
BSX47	BFR21, BFT39, BSY45-55, 40360, 2N1893-3020-3419-4239
BSX48	BFY67C, BSV64, BSX73, BSY51-53, ME6001, MPS3705, SK3122, 2N2221/2-3/05-3736-3866-4013-5450
BSX49	BFX17, BFY67A, BSX74, BSY52-54, ME6002, SK3122, 2N2194-2222/A-2224-2270-2868-3053-4013
BSX51	BCY58/VII/A, BFY76, BSW62, BSX38-53A-69-81A, SE1001, SK3122, TIS86, 40398, 2N697-929-2222-3710
BSX51A	BC107A, BCY59A, BSX54A, SE4020, SFT714A, SK3122, TIS87, 40407, 2N929A-2222
BSX52	BCY58/VII/B, BFY77, BSX53B-67-79, SE1001, SK3122, TIS86, 40397, 2N930-1420-2222-3711
BSX52A	BC107B, BCY59B, BFX93A, BSX54B, SE4020, SK3122, TIS87, 40407, 2N330A-929A-2222A
BSX53	BCY58/VIII, BSX51, SK3122, 2SC907, 2N1613-2222
BSX53A	BC208A, BCY58A, BSX51, SE1001, TIS86, 40398
BSX53B	BC208B, BCY58B, BSX52, SE1001, TIS86, 40397, 2N930-3711
BSX54	BC107A, BCY59, SK3122, 2N2222/A
BSX54A	BC208A, BCY59A, BSX51A, SE1001, TIS86, 40398, 2N920-3710

BSX54B	BC208B, BCY59B, BSX52A, SE1001, TIS86, 40397, 2N930-3711
BSX59	BFX69, BFY99, BSX45, 2SC479, 2N2218A-2787-3252-3444-3553-3725-3735-5262
BSX60/I	BFX69A, BFY99, BSX45, SK3024, 2SC479, 2N2219A-2444-2787-3444-3553-3735-5262
BSX62	BFX34, BFY51, BSW65, SK3024, 2N1889
BSX62VI	BFW47, BFX17, BSX23, 40348, 2N2194A-2788-2831-2890-3506-3735-3831-5531
BSX62X	BFW47, BFX69, 40348, 2N2193A-2788-2891-3506-3553-3735-5531
BSX62XVI	BFW47, BFX68, 40348, 2N2192A-2789-2891-3506-3553-3735-5531
BSX63	BFX34, BSW65, SK3024, 2N1893
BSX63VI	BFY69A, BSX23, 40348- 2N2890-3109-3507-3553-3830-5531
BSX63X	BFY68A, 40348, 2N2891-3109-3507-3553-5531
BSX67	BC108A, BCY58/A, BF154, BSX38A-51-69, SK3122, 2N743-2708-3691-3709-4874-5135
BSX68	BC148-168A-238A, BFY39, BSW88A, SK3122, 2SC641, 2N2708-2884-3691-4874-5135
BSX69	BC148-169A-239A, BFY39, BSX54A, BSW42, SK3122, 2N2708-2884-3692-4124-4874
BSX70	BFY34, BSX33, BSY44-53, SK3024, 2N718A-1613-2195A-2297-2790-3253-3735-3831
BSX71	BFY46, BSX33, BSY54, SK3024, 2N1711-2195A-2222-2297-2791-3252-3727-3831
BSX72	BC140, BFX96, BSX48-70, BSY51-58-81, MM8001, MM3724, SK3024, 2SC479, 2N2195-2219-3724/A
BSX73	BFX96, BFY34, BSW51-84, 2N2218-2410-2537-3299
BSX74	BFX97, BFY46, BSW52-85, 2N2219-2538-3300
BSX75	BC140, BSW41-61-82, BSX20-48, BSY51, MM3724, SK3122, 2N2195-2222-3704/A-4013
BSX76	BC238A, BSW88A, BSX51, BSY62-71, SK3122, 2N835-2369-3014-3298-3394-4124-5136
BSX77	BC237A, BSX26-51A, BSY63-76, SK3122, 2N2369-2601-3014-3227-3566-4072-4137
BSX78	BC237B, BSX26-51A-79, BSY75, SK3122, 2N2369-3014-2501-3227-3566-4072-4137
BSX79	BC237A, BCY59VII, BSX51A, BFX17, BSY74, SK3122, 2N1983-2219-2222-2846-3261-3705-3946
BSX80	BFX45-89, BSX28-68- BSY19-53-63, NKT13429, SK3122, 2SC641, 2N708-2708-3011-3691
BSX81	BC148A-183A-208A-238A, BCY58, BFX45, BSW33-88, BSY74, MPS6574, NKT10419, SE1001, SK3122, 2N718-2410-3565
BSX87	BFX60, BSX20-75, BSY81, ME6002, NKT13329, SK3122, 2N914-3009-3210-3646-4427
BSX87A	BFX60, BSX19-72, BSY80, ME6001, NKT13429, SK3039, 2N708-914-3013-3211-3647-4427
BSX88/A	BSY19-63, NKT13329, SK3039, 2N708-3013-3211-3648-4427
BSX89	BSY21-62, SK3122, 2N706A-2369A-3303-3426-4138

BSX90	BFY66, BSY17-38, NKT13329, SK3122, 2N743-3011-5187
BSX91	BFY66, BSY18-39, NKT13329, SK3122, 2N744-918-3011-5187
BSX92	BSX19-48-72, BSY81, ME6001, NKT 329, SK3039, 2N914-2368-3009-3210-3646
BSX93	BSX20-48-75, BSY82, ME6002, NKT13429, SK3039, 2N914-2369-3013-3211-3647
BSX94	BFX12, BSW19VI-81, ME8201, 2N2894A-3304-3639-4207-4257-5140/1
BSX95	BSX33-45/VI, BSY44-53, ME6001, SK3024, 2SC479, 2N698-2195A-2297-2790-3253-3831
BSX96	BSX33-45/XVI, BSY54-71, ME8003, SK3024, 2SC479, 2N699-2195A-2297-2790-3252-3735-3831
BSX97	BFX60-96, BSX20-48, BSY51-81-91, ME6001, MM3724, SK3122, 2N2195-2218-3724A
BSY10	BC141VI, BCY65E, BF177, BFX92A, BSY44/5, ME4101, SE4020, SK3020, 2N696-735-929A-2218A-2222-2483
BSY11	BC140VI, BCY59/VII-66, BFX92, BSY44, ME4102, SK3020, 40240, 2N736-929-2218-2222-2484-3693
BSY17	BSX90, BSY19-21-38-70-95, NKT13329, SK3122, 2SC321, 2N743-914-2368-3011-5187
BSY18	BCY16-70, BSX20-91, BSY21-39-95A, NKT13329, SK3122, 2N744-2368-3011-5187, 2SC321
BSY19	BSX20-88, BSY63, SK3122, 2SC321, 2N708-2242-4427-4873
BSY20	BSX68-80-89, BSY18-62-70-73, MPS6579, SK3122, 2N706/A/B-3572-4264-5187
BSY21	BSX20-87, BSY10-19-63, SK3122, 40405, 2N914-3013-3210-3648
BSY22	BF115, BFY39-74, BSX38, BSY63, MPS916, SK3039, 2N708-916-3014-3933-4427-4876
BSY23	BFY39, BSX38, BSY22-63, MPS834, SK3039, 2N708-834-2368-3014-3932
BSY24	BSX19-48-62-72, BSY81-91, ME6001, 2N2218/9-2368-3508-3724A-3866-4046
BSY25	BSX20-48-62-72-75, BSY26-46-58-82, ME6002, SK3024, 2N2218/9-2369-3508-3724A-3866-4046
BSY26/7	BFY66, BSY17/8/9-38/9-95, ME9001, NKT13329, SK3122, 2N706A-708-743-914-3011-5187
BSY28	BFY90, BSW78, BSX91, BSY17-19-21-61, ME9002, NKT13329, SK3122, 2N708-743-914-2369-3011-4264-5128-5187
BSY29	BFY90, BSW79, BSX70-91, BSY18-21-39-62, ME9002, SK3122, NKT13329, 2N708-744-914-2369-3011-4264-5129-5187
BSY32	BFY66, BSX70-91, BSY17-21-29-38-95, ME9001, NKT13329, 2N743-3011-5187
BSY33	BFY66, BSX70-91, BSY18-21-29-39-95A, ME9001, NKT13329, 2N744-3011-5187
BSY34	BFX69, BFY44, BSW53, BSX61-73, BSY44, SK3122, 2SC479, 2N1613-2218/A-2219A-2410-3252-3724

BSY36	BFY90, BSX91, BSY17/8-21-38/9-70-95A, ME9001 NKT13329, 2N708-743/4-3011
BSY37/8	BFY90, BSX91, BSY17-21-38-70-95, ME9001, NKT13329, 2N708-743-3011-5187, SK3122
BSY39	BFY66, BSX91, BSY17/8/9-21-95, ME9001, NKT13329, 2N708-744-3011-5187
BSY40/1	BCY78, BFX48, BSW19VI-21A, BSX40, BSY59, SK3114, 40406, 2SA548, 2N2297-2411-2894-3638-3829-4126-5040-5138
BSY42/3	BFX73, BFY12, 2N2219
BSY44	BFX69, BFY34-53-67, BSX45-95, BSY53, ME6002, SK3024, 2SC708A, 2N1613-3678-3735-5321
BSY45	BC141VI, BFR21, BSX45-47, BSY55, ME8002, SK3024, 2SC708A, 2N739-1893-2243A-2405-3565
BSY46	BFY44-55, BSX39-45-46/VI, BSY84, ME8003, SK3024, 2SC708A, 2N1613-1893-2193-2297-3108-3507-3722
BSY47/8/50	BSY20-62-70, 2N706A
BSY51	BFX55, BFY34, BSX45/VI-60-73-95, BSY51, SK3024, 2SC479, 2N697-2410-3015-3299
BSY52	BFX55, BFY34-56, BSX45/X-74-96, SK3024, 2SC479, 2N1420-1711-2219-2538-3053-3301
BSY53	BFX69, BFY34-67, BSX45-46VI, BSY44, SK3024, 2SC479, 2N1613-2788-3678-3735-5262
BSY54	BFX68, BFY46-68, BSX45-46X-47, BSY71, BFR21, ME8002, SK3024, 2N1711-2049-2219A-2789-5321
BSY55	BFR21, BFY14C, BSX46VI-47, BSY45, ME8002, SK3024, 2N1893-2102-2243A-2895-3498-3566
BSY56	BFR21, BFY14D, BSX46X-47, BSY55, ME8002, SK3024, 2N739-1890-1893-2102-2243A-3565
BSY58	BFX17, BFY44, BSX73-95, BSY51, SK3122, 2SC479, 2N697-3705-3734-3866-4013-5450
BSY59	BC328, BFX48-88, BSV15VI, BSX40, BSW44A, ME0402, SK3122, 2N1132-2838-3135-3702-4037
BSY60	2N706A
BSY61	BC238A, BSX27, BSY19-21, ME9002, SK3122, 2N706A-753-835-2205-3011-5183
BSY62	BSX20-89, BSY19-20-21-70, ME9002, SK3019, 2N706/A-708/A-3303-3426-5065-5183
BSY62A/B	BSX20, BSY62-70, 2SC321, 2N706A
BSY63	BSX20-88, BSY19, ME9001, SK3122, 2SC321, 2N708-2195-2369-3137-3706-4427-4875
BSY68	BF111, BFS47, BSV64, BSW10, BFY65, BSY55, ME8002, MM3000, 2N720A-1990-2443-2509-3114-5184
BSY70	BSX24-48-76-87, BSY19-62, ME9002, MPS706, SK3039, 2SC321, 2N706-753-2205-3563-3839
BSY71	BFY46-68, BSX45/XVI, BSY54, ME6102, SK3024, 2N1711-2049-2219/A-2789-5321

BSY72	BC108/A-130A-139-182A-183-207A, BCY58VII, BSY19-63, ME9002, MPS6520, SK3019, 2SC648, NKT13429, BSX5-28, 2N930-2330-2708-3014-3932-4874-5136
BSY73	BC108/A-130A-182A-183-207A, BCY58VII, BSX29-51-53, BSY19-63, ME9002, MPS6520, NKT13329, SK3019, 2SC907, 2N2221-2331-2708-3014-3932-4874-5136
BSY74	BC108/A-130A-182A-183-207A, BCY58VIII, RFY46-67, BSX20-48-53, BSY53-67, NKT13429, SK3019, 2SC907, MPS6520, 2N929-2331-2708-3014-3932-4874-5136
BSY75	BC107-182, BCY59VII, BFX51A-79, BFY33-68, BSX29-48-51A, BSY54-71, MPS834, SE6001, SK3122, 2BC907, 2N834-2221/2-3691-3933-4876
BSY76	BC107-182, BCY59VII, BFY12C-68, BSX29-45-48-51A, BSY54-71, SE6002, SK3122, 2SC907, 2N834-2221/2-3691-3933-4123-4876
BSY76B	BC108A
BSY77	BCY65/E, BFR19, BFY34, BSX21-45-47, BSY55/6, ME8002, SK3024, TIS98, 40408, 2N735-2219-222/A/2A-2297-3722-4961
BSY78	BCY65, BFR20, BFY46, BSX45-47, BSY45-55-78, ME8002, SK3024, TIS98, 40409, 2N736-2222/A-2297-3019-4945
BSY79	BFX98, BFY45-65-80-90, BSX21-45, BSY68, ME1120, SE7016, SK3044, 2SC857, 2N1990-2258-3114-5184
BSY80	BC108A-109-130C 184-209, BCY58VIII, BFX89, BFY66, BSX28-51, BSY62, MPS6520, NKT13329, SK3019, 2SC907, 2N918-930-2331-2708-3932-4874-5136
BSY81	BFS95, BFY52-67C, BSX32-45VI-72, BSY58, SK3024, 40407, 2SC708, 2N2193-2195-2218-3110-3724-4013
BSY82	BFS95, BFY52-67C, BSX32-45X-48-72-75-95, SK3024, 40407, 2N2192-2195-2219-3109-3724-4013
BSY83	BFY55, BSX32-45VI, BSY40-46, ME6101, SK3024, 40348, 2SC708, 2N1613-2193-2218A-2297-2351-3110-3725
BSY84	BSX32-45X, BSW85, ME8003, SK3024, 40348, 2N2192/3-2219/A-2297-2351-3109-3725
BSY85	BFR21, BSX46VI, BSY46-55, ME8002, SK3024, 2S708A, 2N1889-1893-2193A-2243A-2405-2509-3108-3565
BSY86	BFR21, BSX46X, BSY56, ME8002, SK3024, 2N1890-1893-2193A-2243A-2405-2509-3019-3107-3566
BSY87	BC141VI, BFY14C, BSX39-46VI, BSY55, ME8002, SK3024, 2N699B-1889-1893-3108
BSY88	BC141X, BFY14D, BSX39-46X, BSY56, ME8002, SK3024, 2N699B-1890-1893-3019-3107

BSY89	BSX28-51, BSY19-63, ME9002, NKT13429, SK3122, 2N2330-2432-2708-3932-4874-5136
BSY90	BC140XVI, BFX55, BFY46, BSX74-96, BSY52, SK3024, 2N1420-1711-2219-2270-2540-3015-3053-3301-3641
BSY91	BFY33, BSX19-25-48-51A-60-72, ME6001, SK3024, 2SC708, 2N916-2218/9-2368-2710-3706-4427-4875
BSY92	BFX55-96, BFY34, BSW51, BSX45VI-73, SK3024, 2SC479, 2N1959-2219/A-2224-3053-4400
BSY93	BCY65E, BFX55-94, BSW61, SK3122, 2N1613-2221/2-2222A-2897-4400
BSY95	BSX51-91, BSY20-62, ME9022, SK3019, 2N743-914-1708-2369-5187
BSY95A	BC108A, BSX51-73-91, BFX96, BSY20-58-62, ME9022, SK3122, 2SC321, 2N744-914-1708-2218-2369-5187
BSY149A	BSX66
BSY149B	BSX67
BSY152A	BSX19
BSY152B	BSX20
BTX18-100	2N1596
BTX18-300	2N1598
BTX18-400	2N1599
BU100	BD130, BUY20, 2N2821-3055-3442-3773
BU102	BU104-106-110, 2N2581-2583-3902-5239
BU103A	2N4912
BU104	BU102-108-111, BUY21, 2N2581-3902-5240
BU105	BUY62, SK3115, 2SC937, 2N4240-5157
BU106	BU102-109-111, 2N2581-3902-5239
BU107	BU102-109-110, 2N3902-5157-5240
BU108	BUY65, SK3115, 2N3285-3902-5157
BU109	BU102-108-111, BUY64, 2N2581-3902-5157-5240
BU110	BU107-109, BUY18, 2N2581-3902-5157-5240
BU111	BU102-106-109-126, 2N2580-3902-5157-5240
BU120	BDY98
BUY10	BD106-136, BUY19, SK3027, 2N1616-1720-3771-5527
BUY11	BD136, BDY15A, BUY19, SK3027, 2N1616-1720-3772-5527
BUY12	BDY19-25-91, BUY16-21, PP3001, 2N2581-3902-5240
BUY13	BDY18-24-91-93, BUY20-24, PP3002, 2SC665, 2N3055
BUY14	BDY17-23-61, BUY19-20, PP3002, 2SD125A, 2N3055-4911
BUY16	BDY54-90, BUY20, PP3002, 2N3055
BUY17	BDY53-60, BUY12-21, PP3003, 2N3055
BUY18	BDY19, BU109-110, BUY23A, 2N2581-3079-3902-5240
BUY19	BD130, BDY17-53, BUY23A, 2N1616-3902-5240-5534-5730
BUY20	BDY25, BU110, BUY18, 2N3079/80-3902-5240

BUY21	BDY26, BU111, BUY18, 2N3080-3902-5240
BUY22	BDY27, BU102-111, 2N3080-3902-5240
BUY24	BDY24-34-60, BUY13-20, 2N3055-5069
BUY26	BDY25, BUY18-20, 2N2581-3080-3442-3902-5239
BUY27	BDY26, BUY18-21, 2N2581-3080-3902-5240
BUY28	BDY27, BU102, BUY22, 2N2583-3080-3902-5240
BUY38	2N3054
BUY43	BDY80A, BUY19-20, 2N1616-3055-5530
BUY46	BDY81, BUY19, 2N1616-3045-3055-5532
BUY51A	2N3772
BUY59	BDY25, BU110, BUY18, 2N2581-3079-3902-5239
BUY60	BDY26, BU102 111, 2N2501-3080-3902-5239
BUY61	BDY27, BU110, BUY18, 2N2581-3079-3902-5240
BUY62/3/4	BDY28, BU102-111, 2N2581-3080-3902-5240
BUY65	BDY28, BU102-111, 2N2581-3902-5240
BUY66/7	BU102-110, 2N2581/2-3738-3902-5240
C94/A/E	E201
C95/A	E201
C95E/96E	E203
C97E/98E	E402
C106	BC283, BCY29-34, SE8540, SK3025, 2N721-1131-4036
C111E	BCY78, BSX19-38-53, BSY21, NKT13329, SE1001, SK3122, 40294, 2N708-743-914
C400	BSW84, BSX33-48-70-75, SK3122, 2N2221-2529-2845-3115-4951-4994-5188
C407	BF117-178-186-257, BFR88, BFX98, MM2258, SE7016, TIS101, 2N3712-5184
C413E	E111
C420	BFY51, BSX32-46VI, BSY51, SK3024, 40347, 2N2192-2194-2217-2410
C424	BFY33-52-72, BSW82, BSY91, SK3122, 2N2195-3252-3724-3866-5145
C425	BFR19, BFY34, BSX46VI-59, BSY53, ME8003, SK3024, 2N2787-3253-3678-3737-5262
C426	BSX32-46VI-60, BSY51-54, SK3024, 40347, 2N2194-2217-2410
C442	BFY33-52-72, BSW82, BSY91, 2N2195-3252-3724-3866-5145
C444	BSW41, BSX29-39-48, MPS3705, SK3122, 2N2221-3705-5108-5144-5450
C450	BC207A-237A-261A-385A, MPS6575, SK3122, 2N929A-3565/6-4242A
C673	E203, SK3112
C674	E204, SK3112
C4241	2N2704
C6690/1	E203
C6692	E202
C9080	BC283-360VI, BCY54-78A, BSV15VI, MPS6516, SE8540, 40362, 2N3644-5365
C9081	BC283-360X, BCY54-78B, BSV15X, MPS6517, SE8541, 2N3644-5366

C9082/4	BC205, BCY78A, EN3502, MPS6519, SK3114, 2N3702-5448
C9083/5	BC205, BCY78B, EN3502, MPS6519, SK3114, 2N3702-5447
CDT1309	SK3009
CDT1310	AD138, ADZ11, AUY29, AT1138A, MP2061, NKT402, DTG110, SFT265, SK3009, TI3028, 2N1146-2869-5134
CDT1311	AD131/50-150, ADZ11, ASZ17-18, AUY21/IV-28, DTG601, MP2061, NKT402, SFT266, SK3009, TI3028, 2N513A-1146A-2869
CDT1312	ADZ12, AT1138B, AUY22-28, DTG602, MP2062, NKT401, SFT267, TI3029, 2N513B-1146B-2870
CDT1313	AD132, AT1138B, AUY22IV-28-34, ASZ15-16-18, DTG603, SFT268, TI3030, 2N513B-1100-1146B-2870
CDT1315	AT1138B, AUY28-34, DTG603, MP2063, NKT401, SFT258, TI3030, 2N513B-1146C-2870
CDT1319	AD138, ADZ11, AT1138A, AUY29, DTG601, MP2061, NKT402, SFT265, SK3009, TI3028, 2N513A-1146-2870
CDT1320	AD138/50, ADZ11, AT1138B, AUY21, DTG602, MP2062, NKT403, SFT266, SK3009, TI3029, 2N513B-1146A-2870
CDT1321	ADZ12, AT1138B, AUY22-28, DTG603, MP2063, NKT401, SFT267, SK3009, TI3030, 2N513B-1146B-2870
CDT1322	AUY28-34, DTG603, SFT268, 2N1100-1146C
CF2386	E203
CK13	SK3005
CK14	SK3005
CK17	SK3005
CK22	SK3004
CK65	SK3004
CK66	SK3004
CK721/2	OC71, SK3004, 2N2429
CK724	2N2429
CK725/27	OC71, SK3004, 2N2429
CK751	AC128, OC72, SK3004, 2N243-1281
CK759/60	AF185, SK3005
CK761	AF185, SK3005
CK762/6/6A	AF185, SK3006
CK790	SK3004
CK791	SK3004
CK793	SK3004
CK870/1	SK3004, 2N2429
CK872	OC72, SK3004, 2N2431
CK878	OC74, SK3004, 2N2431
CK882/8	OC72, SK3004, 2N2431
CK896A	OC57/8
CK897A	OC58
CK898	OC59
CK898A	OC59-77
CM601/2	E112

CM603	E111
CM640/1	E113
CM642	E112
CM643	E113
CM644	E112
CM645/6	E111
CM647	E110
CM697	E109
CMX740	E105
CP650/1	E108
CP652	E106
CP653	E109
CQT940/A/B	ADY26, DTG2200, MP1531, NKT403, 2N1361-1541-2140-3314
CQT1075	ADY26, DTG2400, MP902, SFT268, 2N908-1533-1543-1905-3618
CQT1076	ADY26, DTG2300, MP901, SFT268, 2N907-1532-1542-1906-3616
CQT1110A	ADY26, DTG2000, MP1529, NKT405, SK3009, 2N1359-1539-2138-3312
CQT1111A	ADY26, DTG2100, MP1530, NKT405, SFT265, SK3014, 2N1360-1540-2139-3313
CQT1112	ADY26, DTG2200, MP1531, NKT403, SFT267, SK3014, 2N1361-1541-2140-3314
CST1773	AD138, ADZ11, AT1138, DTG600, MP1529, NKT451, SFT265, TI156, 40022, 2N277-1146, AUY21
CST1773A	AD138, ADZ11, AT1138, AUY21, DTG601, MP1530, NKT451, SFT266, TI158, 40050, 2N278-1146A
CST1773B	AD138/50, ADZ12, AT1138A, AUY22, DTG602, MP1531, NKT405, SFT267, TI158A, 40051, 2N174-1146B
CST1789	ASZ15, AT1138B, AUY34, DTG603, MP1532, NKT403, SFT268, TI161, 40421, 2N375-1146C
CTP1003	ASZ15, OC28
CTP1004	AD149, OC26, 2N2836
CTP1005	AD149, OC26/7, 2N2836
CTP1006	ASZ16, OC29
CTP1032/3/4	AC125/6, 2N2429
CTP1035	AC125/6, 2N2429
CTP1036	AC126, 2N2429
CTP1104	AD130-149, OC30, SK3009, 2N2062A-2836
CTP1108	AD130/III-149, ADY26, AT1138B, AUY22-28, DTG602, MP1551, SFT267, SK3009, OC26, 2N513B-1146B-2061A-2836
CTP1109	AD130III-150/IV-149, ADY26, AT1138B, AUY22-28, MP1551, SFT266, SK3009, OC26, TI156, 2N513B-514B-1146B-1906-2061A
CTP1111	AD131-132/II/III, ADY26, AT1138A, AUY21-28, ASZ15, DTG600, MP1550, SFT265, SK3009, 2SB472, 2N513A-514A-1146A-1905-2065A
CTP1320/30	AC125/6, 2N2429

147

CTP1340/50	AC125/6, 2N2429
CTP1360	AC126, 2N2429
CTP1390/1400	AF127-185, OC45
CTP1410	AF126-185, OC44
CTP1500	AD138, ADZ12, ASZ15-16, AT1138B, AUY21-22III-34, DTG603, MP1556, NKT401, SFT268, SK3009, 2N1100-1146C-1542-1556-2146
CTP1503	AD104-138, ADY26, ADZ12, ASZ16, AT1138A, AUY22/III-28, DTG602, MP1555, NKT403, SFT267, SK3009, 2N1100-1146B-1541-1555-2145
CTP1504	AD133-138-153III, ADY26, ASZ18, AT1138A, AUY21III-22-28, DTG602, MP1555, NKT402, SFT266, SK3009, 2N1146A-1541-1554-2139
CTP1508	AD133/III-136-138, ADY26, ADZ11, AT1138, AUY21-28, DTG600, MP1554, NKT404, SFT265, SK3009, 2N1146A-1540-1553-2138
CTP1509	ADZ12, SK3012
CTP1514	AD149, OC26, SK3009, 2N2836
CTP1544	ASZ15, AT1138B, AUY34, DTG603, MP1556, NKT401, SFT268, SK3009, 2N1146C-1542-2146
CTP1551	ASZ15, SK3009
CTP1552	ADY26, AT1138A, AUY22-28, DTG602, MP1555, NKT403, SFT267, SK3009, 2N1146B-1541-2145
CTP3500	ADZ12, ASZ15, AT1138B, AUY34, DTG603, MP1556, NKT403, SFT268, SK3009, 2N1146C-1542-2146
CTP3503	ADY26, ADZ12, AT1138A, AUY22-28, DTG602, MP1555, NKT402, SFT267, SK3009, 2N1146B-1541-2145
CTP3504	ADY26, AT1138A, AUY21-28, DTG601, MP1554, NKT402, SFT266, SK3009, 2N1146A-1540-2144
CTP3508	ADY26, AT1138, AUY21-28, DTG600, MP1553, NKT403, SFT265, SK3009, 2N1146-1539-2143
CTP3544/5	ADY26, AT1138B, AUY21-28, DTG601, MP1558, NKT402, SFT266, SK3009, 2N1146A-1542-2145
CTP3552/3	ADY26, AT1138A, AUY21-28, DTG600, MP1557, NKT403, SFT265, SK3009, 2N1146-1540-2143
CV2339	GET114, OC71, CXT1, V10/30A
CV2400	GET106, OC71, V10/30A
CV5105	OC45
CV5309	V15/20P
CV5322	TS10
CV5327	V30/201P
CV5328	2xOC35
CV5329	2xOC28
CV5330	V10/50A
CV5335	SA495
CV5353	XS101
CV5359	2S001
CV5396	3S004
CV5411	2xGET110
CV5457	OC70
CV5459	GET104
CV5774	GT35

CV5776	2N144
CV5779	CTP1245
CV5780	CDT1339
CV5791	V30/20DP
CV5792	2S702
CV5814	TK30C
CV5860	ZT23
CV5861	2G401
CV5863	ZT21
CV5885	TK42C
CV5909	2S005
CV5910	2S019
CV5929	V60/201P
CV5949	2G220
CV5967	2G403
CV5974	2S102
CV5975	2S103
CV5994	2G225
CV5995	2S703
CV7001	ACY31, GET103, 2G303
CV7002	GET116, 2G382
CV7003	ASY55, OC44, 2G302
CV7004	ASY54, GET873, OC45
CV7005	GET103, OC71, 2G371, 2N280
CV7006	GET103, OC72, V10/50A, 2N281-374
CV7007	GET111, 2G377, OC77
CV7008	GET106, V10/50A
CV7009	GET110, V60/201P
CV7010	OC26, V30/30P, 2N1314
CV7011/2	V60/30P
CV7042	ASY63, TK29B, 2G302
CV7043	BC210, 2S322
CV7044	BCZ11, 2S323
CV7056	2S002
CV7057	2S003
CV7058	2S004
CV7059	2S005
CV7060	2S014
CV7061	2S012A
CV7062	2S017
CV7063	2S018
CV7064	2S019
CV7065	2S020
CV7066	2S013A
CV7074	GET103
CV7075	OC201, 2S323
CV7080	GET571
CV7081	GET572
CV7082	GET573
CV7083	NKT402, OC29
CV7084	NKT404, OC35
CV7085	NKT401
CV7087	GET875, 2G306
CV7089	OC171, 2G401

CV7107	ACY31
CV7115	V30/30DP
CV7116	V60/30DP
CV7118	XB114
CV7129	OCP71
CV7131	ZT20, 2N1564
CV7132	ZT21, 2N1565
CV7133	ZT22
CV7134	ZT23, 2N1564
CV7149	2S302
CV7151	2S303
CV7152	2S301
CV7154	ZT43, 2S103
CV7185	ASY66
CV7186	ZT45
CV7187	ZT46
CV7326	2G377
CV7327	2G371
CV7328	2S104
CV7333	2S025
CV7334	2S026
CV7336	2N743
CV7337	2S131, 2N744
CV7339	SAC42B
CV7340	SAC40
CV7341	2S300
CV7342	2S300
CV7343/4/5	2S300
CV7346	2S300
CV7348	2N1302
CV7349	2N1304
CV7350	2N1306
CV7351	2N1308
CV7352	2N1303
CV7353	2N1305
CV7354	2N1307
CV7355	2N1309
CV7361	BUY11
CV7362	BSY25, 2N2217
CV7363	2S305
CV7366	2S305
CV7371	2N2220
CV7372	2N2221
CV7379	2N916
CV7390	2G102
CV7391	2G110
CV7393	2N705
CV7394	2N711
CV7396	2S307
CV7404	2N1893
CV7430	BSY26
CV7431	BSY27, 2N753
CV7440	2N1613

CV7448	ASY60
CV7459	2S300
CV7461	2S103
CV7462	2N2411
CV7463	2N2412
CV7464	2N706A
CV7477	C63
CV7478	2N918
CV7479	2N2060
CV7484	2H1254
CV7485	2H1255
CV7486	2H1256
CV7487	2H1257
CV7488	2H1258
CV7489	2H1259
CV7490	BFY17
CV7491	BFY18, 2N929
CV7492	2N929
CV7493	2N930
CV7495	2N696
CV7496	2N697
CV7497	C64
CV7527	ZT1487
CV7528	ZT1488
CV7529	ZT1489
CV7530	ZT1490
CV7554	2N2475
CV7555	2N2369A
CV7580	2N1131
CV7581	2N1132
CV7588	2S104P
CV7596	2S103
CV7598	2S301
CV7602	ZT152
CV7603	ZT153
CV7604	ZT154
CV7629	2S024
CV7630	2N2696
CV7631	2N2904
CV7632	ZT92
CV7633	ZT92
CV7637	D1003
CV7638	D1004
CV7639	ZT90
CV7644	ZT89
CV7647	2S322
CV7648	BSY95A
CV7750	ZT81
CV7751	ZT181
CV7752	ZT82
CV7753	ZT182
CV7754	ZT87
CV7755	ZT187

CV7763	2N2218
CV7764	2N2219
CV7765	2N2218A
CV7766	2N2219A
CV7767	2N2221
CV7768	2N2222
CV7769	2N2221A
CV7770	2N2222A
CV7774	2N2194
CV7775	2N2243A
CV7779	ZT7400
CV7832	VX9260
CV7837	2N2894
CV7854	2N3055
CV7855	2N3821
CV7856	2N3822
CV7857	2N3823
CV7858	2N3824
CV7867	BDY52
CV7868	BFX18
CV7869	BFX19
CV7870	BFX20
CV7871	BFX21
CV7872	BFX18
CV7873	ZT600
CV8015	2S018
CV8276	2×V30/30P
CV8288	TK13
CV8322	SG220
CV8335	2×2G210
CV8358	MA393
CV8359	2G306
CV8360	ZT23
CV8361	2G104
CV8368	2G221
CV8369	2G140
CV8372	2G304
CV8373	2S108
CV8385	ZT83
CV8386	V15/201P
CV8393	2S301
CV8400	2G220
CV8414	DT1121
CV8415	DT1111
CV8416	DT1112
CV8424	DT4111
CV8442	SAC40
CV8451	2S019
CV8453	2S002
CV8455	2S003
CV8473	2×2S324
CV8500	2S321
CV8562	TS774

CV8582	GT42
CV8611	2S024
CV8612	2S323
CV8613	2S712
CV8615	BSY95A
CV8643	2S701
CV8647	ZT92
CV8649	ZT86
CV8673	2G227
CV8677	2N2270
CV8686	2xNKT401
CV8696	2S711
CV8702	2N976
CV8722	2S306
CV8762	2S322
CV8781	2S020
CV8782	2S026
CV8783	2S304
CV8789	2x2N458A
CV8792	CTP1109
CV8806	ZT1481
CV8818	2S722
CV8873	V205
CV8887	2x2S017
CV8910	2S025
CV8913	2N2218
CV8942	2S303
CV8967	MA520
CV8970	2S302
CV9002	2S320
CV9035	2S721
CV9052	C444
CV9053	MM2712
CV9058	2S324
CV9100	40290
CV9101	2N2040
CV9105	GT46
CV9111	GT43
CV9114	2S732
CV9177	2G403
CV9183	TS13
CV9187	ST701
CV9190	NKT271
CV9195	2N3055
CV9209	2S018
CV9217	MM1614
CV9218	40251
CV9220	SA56
CV9243	40110
CV9249	2G401
CV9258	2XGET103
CV9321	2S104
CV9347	SE3001

CV9392	ZT2475
CV9415	2G302
CV9449	2x2N3858
CV9490	TI3028
CV9505	T2039
CV9506	T2775
CV9516	ZT211
CV9525	T2040
CV9530	2xT2040
CV9565	2S733
CV9566	2S305
CV9589	40250
CV9590	40231
CV9603	2N2892
CV9625	MM2614
CV9626	MM1613
CV9627	MM2613
CV9629	V405
CV9760	ZT1483
CV9761	ZT1484
CV9762	ZT1485
CV9763	ZT1486
CV9823	CTP1500
CV9861	ZT1479
CV9862	ZT1480
CV9863	ZT1481
CV9864	ZT1482
CV9865	ZT1487
CV9866	ZT1488
CV9867	ZT1489
CV9868	ZT1490
CV9888	ZT87
CV9959	ASY60
CV9977	2S033
CV9978	2S034
CV9979	2S035
CV9980.	2S036
CV10040	2S731
CV10053	ZFT14
CV10082	2S104
CV10113	MA393C
CV10150	ZT81
CV10153	SAC42A
CV10161	ZDT21
CV10164	2G387
CV10165	ST95A
CV10166	40321
CV10180	2S325
CV10181	2S720
CV10195	SA52
CV10240	ZT1514
CV10244	ZT1513
CV10279	ZT1072

CV10292	2G303
CV10394	2G382
CV10396	2S502
CV10397	SAC44
CV10409	2S327
CV10440	BC107
CV10467	NKT12
CV10576	2G309
CV10692	2S024
CV10697	2N491
CV10715	40354
CV10718	OC36
CV10739	ZT1708
CV10740	ZT183
CV10741	ZT184
CV10750	ZT90
CV10751	2G386
CV10834	2S501
CV10845	2S033
CV10862	ZT81
CV10863	ZT403P
CV10866	ZT42
CV10879	2S300
CV10913	TI3031
CV11060	ZT88
CV11070	2S024
CV11071	2S025
CV11072	2S026
CV11089	ZT152
CV11091	ZT210
CV11092	ZDT42
CV11093	ZDT41
CV11096	ZT91
CV11145	40346
CV11152	2S512
CV11164	ZT114
CV11258	ZT44
D16P1/3	MPS-A13
D16P2/4	MPS-A14
D28A5/6/12	MPS-U05, SK3041
D28A13	MPS-U05, SK3041
D28D1/2	MPS-U01, SK3054
D28D4	MPS-U02, SK3054
D29A4	SK3118, 2N3702-4125
D29A5	SK3114, 2N3702-4126
D29E1/2	SK3114, MPS6562
D29E4/5/6/7	MPS-A55
D29E8/9/10	MPS-A56
D33D21	MPS6560, SK3122
D33D24/5/0/7	MPS-A05
D33D28/9/30	MPS-A06
D40D1/2/3	SK3054, TIP29
D40D4/5/7/8	SK3054, TIP29A

D40N1/3	MPS-U10
D41D1	MPS-U51, TIP30
D41D2	MPS-U51, TIP30, SK3083
D41D4/5/7/8	TIP30A
D42C1/2/3	SK3083, TIP31
D42C4/5/7/8	SK3083, TIP31A
D43C1/2/3	TIP32
D43C4/5	TIP32A
D43C7	SK3083, TIP32A
D43C8	TIP32A
DR100	SK3016, TS1-14
DR101	TS1
DR102	TS2-14
DR108	TS1
DR109	TS1-14
DR110	TS3
DR126	OC66, TS1-14, OC66
DR127	TS2-14
DR128	OC65-66, 2N465
DR129	TS2-14, 2N465
DR130/1	TS1
DS25	AF185, SK3008
DS26	AC128, SK3004, 2N2431
DS34/41	AF178, SK3006, 2N2495
DS44	AC127, SK3010, 2N2430
DS246	AC126, 2N2429
DT80	CTP1502, SK3010, 2N677C-1031B
DT100	CTP1500, SK3010, 2N677C-1031C
DT1003	BFY44, SK3045
DT1111	SK3024, ZT1479
DT1112	SK3024, ZT1480
DT1121	SK3024, ZT1481
DT1122	BDY11, SK3024, ZT1482
DT1511	SK3024, ZT1700
DT1521	SK3024, ZT2270
DT4110	SK3027, BDY10
DT4111	SK3027, ZT1488
DT4121	SK3027, ZT1490
DTG110	ADY26, AUY22-28, MP1612A, NKT403, SFT267, SK3009, 2N1146-1541-2145
DTG110A	ADY26, AUY28-34, MP1612A, NKT402, SFT268, 2N1146A-1542-2145
DTG110B	ADY26, AUY21-28, MP1612, NKT401, SFT265, SK3009, 2N1146-1540-2144
DTG600/1	ADY26, AUY22-28, MP1612, NKT402, SFT265, 2N1146-1541-2145
DTG602	ADY26, AUY28-34, MP1612A, NKT402, SFT267, 2N1146B-1543-2870
DTG603	ADY26, AU106, AUY28-34, MP1612B, SFT268, 40440, 2N1146C-5325
DTG1010	AU107, MP3731, 40439, 2N5324
DTG1110	AU108, MP3730, 40439, 2N908
DTG2100	ADY26, AUY22-28, MP2200, NKT403, 2N1539-2143

DTG2200	MP3618, NKT403, 2N1100-1540-2144
DTG2300	MP2300, NKT403, SDT1950, 2N1100-1542-2146
DTG2400	MP2400, NKT402, SDT1960, 2N1100-1543-2870
DTS401	BDY28, BU102-111, BUY67, 2N2580-3902-5241
DTS402	BDY28, BU102-111, BUY66, SK3111, 2N2581-3902-5241
DTS410	BU110, BUY18-20, 2N2822-3080-3902-5240
DTS411	BU102-104-110, BUY67, 2N3902-5241
DTS413	BU102-104-111, BUY67, 2N2582-3902
DTS423	BDY28, BU102-111, BUY67, 2N2582-3902
DTS424	BDY28, BU102, BUY67, SK3111, 2N2583-3902
DTS425	BDY27, BU102, BUY67, SK3111, 2N2583-3902
DTS430	BDY27, BU102-111, BUY67, 2N2582-3902
DU4340	E402
DW6208	BC107B-129B-182B, MPS6566
DW6577	BC107/A-135-182, MPS6566
DW6737	BC107A, MPS6566
DW7000	BF167-225
DW7035	BC107B-113-167B-171B-182/B, MPS6566
DW7039	BF110-117-156-178
DW7050	BF176-223/4-232
E100	E203
E101	E201
E102	E201, SK3016
E103	E203, SK3112
EN697	BFY34-67, BSX30, BSY51-92, ME6101, SK3122, 2N2410
EN706	BSV89, BSW82, BSX19, BSY62-70, ME6003, NKT13329, SK3122, 2N2369-5276
EN708	BSV92, BSY19-63, ME9001, NKT13429, SK3122, 2N708-2368-5263
EN718A	BFY34, BSX33-61, BSY44-53, ME6001, 2N718A-1613-2787
EN722	BCY70, BFX74, BSV15VI, BSW74, ME0401, SK3118, 40362, 2N1132-2907-3702-5448
EN744	BFY39, BSV90, BSX72, BSY18-95A, ME3011, 2N744-3011-3839-4264-5187
EN870	BFR21, BFY14C, BSV64, BSW10, BSY87, ME8002, 2N1889-2102-2243-3108-4239
EN871	BFR21, BFY14D, BSX46XVI, BSY88, ME8003, 2N1890-2405-3057-3107-3499
EN914	BSX20-26, BSY21-63, ME9001, 2N914-2369-3210-3261
EN915	BFX94, BFY34-67, BSW84, BSY53-93, ME4101, 2N915-2221A-2790
EN916	BFW68, BFX60, BSW41-82, BSX75, ME9001, 2N916-2432A-2708-3211-4873
EN918	BFX59-73, BFY66, BSX19, ME8002, 2N918-2857-3478-3600-3839
EN930	BCY56-59D-66, BFY77, BSX54B, GI3694, 2N930-3691-4424
EN956	BFR20, BFY46-68, BSY54-71, ME8003, 2N1711-2405-3109-3559

EN1132	BCY70, BFX39, BSV16VI, BSW74, ME0401, 40406, 2N1132-2800-2904-3133/4
EN1613	BFY34-67, BSX30, BSY44-53, ME6101, 2N1613-2790-5321
EN1711	BFX68, BFY46-68, BSY54-71, ME8003, 2N1711-2049-3109-3569
EN2219	BFX68, BFY46, BSW52, BSY44-54, ME6102, 2N2219-3300-4227-4653
EN2222	BFX96, BFY46, BSW62, BSY54-71, ME6102, 2N2222-3300-4227-4401
EN2369A	BSV91, BSX20, BSY21-63, ME9001, 2N2369A-3210-3646-4275
EN2484	BCY65X, BC341XVI, BFR16, BFX29, BSX52A-54B, TIS98, 2N2484-3117
EN2894	BFX12, BSW19VI-21, BSX94, ME8201, TIS53, 2N2894/A-4258-4313
EN2905	BFY64, BSV16VI, BSW75, 2N2905-3485A/6A-4028-4037
EN2907	BC360XVI, BSV16XVI, BSX36, SK3118, 2N2907-3486A-3505-4037
EN3009	BSV92, BSY19-63, ME9001, SK3122, 2N3009-3646-4427
EN3011	BFX59, BFY39-66, BSV89, ME9001, 2N2369-2708-3011-3646
EN3013	BSV92, BSY19-63, ME9002, SK3122, 2N703-2708-3013-3646
EN3014	BSV92, BSY19-63, ME9001, SK3122, 2N703-3014-3646-4427
EN3250	BFX48-88, BSV15VI, BSW41A-75, ME0402, 2N3250-4037
EN3502	BC327-360XVI, BSV16XVI, BSX36, 2N2905-3502-4037-5366
EN3504	BC327, BSV16XVI, BSW74, BSX36, SK3118, 2N2907-3504-4037-5366
FE0654A	E212
FE0654B	E211
FE0654C	E210
FE0655A	E111
FE0655B	E112
FE0655C	E113
FM870	BFR21, BFX84, BFY14C, BSW10, BSY87, ME8002, SK3039, 2N1889-2243A-2405-3108-4239
FM871	BFR21, BFX84, BFY14D, BSX46XVI, BSY88, ME8003, SK3039, 2N1890-2405-3057-3107-3499
FM910	BFY14B-57, BSV64, BSW10, BSY87, ME1120, SK3039, 2N1889-1893-1973-2405
FM911	BFR21, BFY14B, BSW67, BSY55, ME8002, SK3039, 2N1890-1893-1974-2405
FM1613	BFX69, BFY34-67, BSY44-53, ME6101, SK3039, 2N1613-2218-2790
FM1711	BFY46-68, BSY54-71, ME8003, SK3039, 2N1711-2219-2405-3109-3569

FM1893	BFR21 BFY45, BSW67, BSY55, ME8002, 2N1890-1893-1973-2405
FM3954A	E413
FM3955A/6	E413
FM3956/8	E414
FP4339	E402
FT107A	BCY59D-65E, BFR17, BSX54B, ME4101, 2N2484-4104
FT107B/C	BC183C, BCY59D-65E/EIX, BSX38B-52A, ME4101, MPS6521, 2N2484
FT709	BFY39, BSV89, BSX72, BSY18-73, ME8101, 2N709-744-3010/1-3493-5128
FT3567	BFR19, BFY84, BSW10-84, BSY53, ME6101, MPS-A05, 2N1613-2193-3567-3725
FT3568	BFX68, BFY46, BSW85, BSX45VI-71, BSY54, ME8003, MPS-A06, 2N1711-3109-3568-4409
FT3569	BFX69, BFY34, BSW84, BSX45XVI-70, BSY53, ME6101, MPS-A05, 2N1613-2193-3569-4047
FT3641/2	BFY34-67, BSW84,BSX45XVI-70,BSY53, ME6101, MPS-A05, 2N2224-2537-2787-2897-3641/2
FT3643	BFY46-68, BSW85, BSX45XVI-71, BSY54, ME8003, MPS-A05, 2N1711-2219-2538-2788-3643
FT3644	BFW20, BC361VI, BFX29, BSV16XVI, ME0402, 2N2905A-3486A-3503-3644
FT3645	BC360X, BFW23, BFX29, BSV15XVI, ME0402, 2N2905A-3486A-3503-3645
FT3722	BFY55-56A, BSW10, BSX45VI, BSY83, ME8002, 40366, 2N2193-2297-3109-3722
FT3820	E175
FT3909	E175
FT4354	BC161VI, BFX29, BFW20, BSV16VI, ME0401, 2N2904A-3485-4030
FT4355	BC161XVI, BFW23, BFX30, BSV16VI, ME0402, 2N2905A/6A-3486-4032
FT4356	BC313, BFX40-41, MPS-A56, 2N4031-4236-4356
FT5040	BC204, BCY78A, BSX36, ME0463, 40406, 2N3467-3502-3905
FT5041	BC204, BCY79A, BSX36, ME0463, MPS-A55, 40406, 2N3467-3502-3905
FTO654A/B	E300
FTO654C/D	E304
FTO654E	E305
FTO655A	E111
FTO655B	E112
FTO655C	E113
GA004	ASY27
GA52829	AC126, 2N2429
GC500	AC121, OC74-302
GC501	AC128, OC79
GC502	AC123, OC80
GC507	AC132, ACY36, OC72-302-307-604sp, TF75
GC508	ACY28-58, OC76-307-308-602spez
GC509	ASY51-52-77, OC77-309II/III
GC510	AC117-128
GC510K	AC128K

GC511	AC188
GC511K	AC188K
GC515	AC125, ACY27-34, OC70-303-601/2, TF65
GC516	AC126, ACY35, OC71-304I-603-604-TF65
GC517	AC116Z-126, ACY23-30-32-33V, ASY58, OC75-304III-350
GC518	ACY33VI, OC75
GC519	ACY33VII, OC75
GC520	AC175-176, SK3004
GC520K	AC176K
GC521	AC187, SK3005
GC521K	AC187K
GCN53	101NU71, 102NU71
GCN54	103NU71
GCN53	GC507-508
GCN56	GC519
GD607/8/9	AD161
GD617/8/9	AD162
GET3	AC126, V10/50A, 2N185-2429
GET4	AC126, NKT228, 2N185, 2429
GET5	AC125, NKT304, V30/201P
GET6	NKT226, TS3, V10/50A, 2N185
GET7	V15/30NP
GET8	V30/30NP
GET9	NKT401
GET15	V15/201P
GET20	V30/201P
GET102	AC156, 2N2429
GET103	AC126, 2N2429
GET104	AC126, ASY59, 2G303, 2N2429
GET105	NKT304, V60/201P
GET106	2N2429
GET110	NKT303, V60/201P
GET111	NKT227, 2G377
GET113	2G301, SK3004
GET114	SK3004, 2G301, 2N2431
GET115	NKT304-351, V15/201P
GET116	NKT304, V30/201P
GET120	NKT303, V30/201P
GET535/6	NKT228
GET538	NKT227
GET571	NKT404, V15/30NP, 2N456/A
GET572	XC141, SK3009, 2N456/A
GET573	NKT403, V60/30NP, 2N457/A
GET574	NKT403
GET581	NKT401
GET582	NKT403
GET583	NKT404
GET584	NKT402
GET706	MPS706, SK3122
GET708	MPS708, SK3122
GET871	SK3006, XA151, 26301
GET872	SK3006, XA152, 2G302

GET873	AF185, SK3006
GET874	AF185, SK3005
GET875	SK3006, 2G306
GET880	AC122-125-136-151; BC260A, BSX36, NKT219, SK3005, 2N1191-1307-1352-1371-1384
GET881	AC117-128-142-153-180, BC260A, BFS12, FT5041, NKT211, SK3005, 2N1305-1384-1924-2000-2374
GET882	AC117-128-142-153-180, BFS12, FT5041, NKT211, SK3005, 2N1305-1925-1998-2000-2375
GET883	AF185, SK3008
GET885	AC117-128-142-153-180, BFS12, FT5041, NKT211, SK3006, 2N1999-2001-2376-3427-4031
GET887	AC122-125-136-151-173, BC260A, BSX36, NKT219, SK3005, 2N1190-1307-1373-1384
GET888	AC122/30-122-126-136-151VII-173, BC260B, BSX36, NKT219, SK3005, 2N1192/3-1307-1309-1375-1998
GET889	AC122-122/30-126-136-151VII-173, BC260B, BSX36, ASY27, NKT219, SK3005, 2N1191-1307-1352-1371-1384
GET890	AC122-122/30-125-136-151VII-173, BC260A, BSX36, NKT219, SK3005, 2N1193-1307-1998
GET891	AC122-125-136-151-173, BC260A, BSX36, NKT219, SK3005, 2G303, 2N1190-1352-1371-1384-5040
GET892/5	AC117-128-142-153-180, BFS12, NKT211, SK3005, 2G304, 2N1198-2374-3424-4030-5041
GET896/7	AC122-125-136-151-173, BC260A, BSX36, NKT219, SK3005, 2N1190-1352-1371-1384-5040
GET898	AC122-125-136-151-173, BC260A, BSX36, NKT141-219, SK3004, 2N1192-1307-1352-1375-1384-5040
GET2369	MPS2369, SK3122
GET3638/A	MPS3638, SK3114
GET3646	MPS3646, SK3122
GF501	AFY11, 2N1141
GF502	AFY10, 2N1142
GF504	AFY10-14, 2N1143
GF505	AF106
GF506	AF106
GF507	AF139
GF514	AF114-124-134
GF515	AF116-126-136
GF516	AF117-127-137
GF517	AF117-127-137
GFT20	AC122R-126-151/IV-125-162-173, BC260A, BSX36, NKT219, 2SB220-459, 2N63-238-1190-1352-1371-2429-5040
GFT20R	AC150-162-172, 2SB32, 2N44
GFT21	AC122/Y-125-126-136-151/V-162-173, BC260A, BSX36, NKT219, 2SB219, 2N266-1191-1352-1373-1384-2429-5040
GFT21/15	AC125, OC604
GFT25	AC122/R-125-126-151IV-163, 2SB101-459, 2N36-238-1190
GFT25/15-30	AC122-125-136-151-171-173, BC260A, BSX36, NKT219, 2N1192-1352-1375-1384-5040

GFT26	AD149, OC26
GFT31	AC126-131-152, 2SB220, 2N44
GFT31/15	AC128
GFT31/60	ACY24, ACZ10, ASY48, 2SB89, 2N24A
GFT32	AC124R-128-132-152IV-153, 2SB222, 2N59-2429-4106
GFT32/15	AC117-128-142-152/IV-184, BFS12, NKT211, 2N1998-2000-2374-3427-4030
GFT32/30	AC117-124-128-142-152/IV-184, BFS12, NKT219, 2N1998-2001-2375-3427-4030
GFT33	AC124R-128-152V, 2SB32, 2N610
GFT33/30	AC117-128-142-152/V-184, BFS12, NKT211, OC308, 2N1925-1998-2375-4030-5041
GFT34	AC106-117-124-128-131-142-153-180, BFS12, NKT211, 2N1999-2001-2375-3427-4031-5041
GFT34/15	AC128-152-153, 2SA219, 2N138
GFT41	AC143, AF106-178-190, BFX48, BSW19-72, NKT603F, 2SA230, 2N2273-2495-3324-4916-5354
GFT42A	AF124-132-134-144-178, BFX48, BSW19-72, NKT674, SFT358, 2SA116, 2N384-1524-2273-2495-3127-4917-5354
GFT42B	AF125-134/5-167-185, BFX48, BSW19-72, NKT674F, 2SA116, 2N384-1525-2189-2412-3323-4916-5354
GFT43	AF125
GFT43/15	AC117-124-128-142-152/V-184, BFS12, NKT219, 2N1998-2000-2375-3427-4031-5041
GFT43A	AF126-135/6/7-146-185, BFX48, BSW19-72, NKT674F, 2SA154, SFT354, 2N310-1526-2190-2411-3324-4917
GFT43B	AF127-136/7-167-185, BFX48, BSW19-72, NKT613F SFT316, 2SA57, 2N1527-1748-2191-2412-4916
GFT43D	AF126
GFT44/5	AF127-132-137-167-185, BFX48, BSW19-72, NKT674F, SK3005, 2N1527-2273-4916-5354
GFT44/15	AF126-127-185, HJ23D, 2N137
GFT2006	AD149, OC16, TI156, 2SB471, 2N456
GFT2006/30	AD1301III-143-149-150-153, ADY24, AUY33, NKT451, OC30, 2SB240, 2N68-178-513-550A-2836-2869
GFT3008/20	AD1301III-131-138/50-142-149, ADY24-28X, ADZ12, MP1530, NKT405, OC30, 2SB86, 2N513A-1530-2148-2836
GFT3008/40	AD131III-138/50-139-148-149, SK3009, TI3029, 2SB86-471, 2N2836
GFT3008/60	AD131-139-148, 2SB86
GFT3008/80	AD132/III-142, ADY23-28, ADZ12, MP1531, NKT403, CTP1111, SFT250, 2SB86, 2N513B-1531-2147
GFT3108/40	AD130/III-132III-143-149-150, ADY23, ASZ15, AUY33, MP1529, NKT451, 2SB472, 2N513-1529-2065-2148
GFT3108/60	AD131/III-138/50-142, ADY24-28X, ADZ12, MP1530, NKT405, 2N514A-1530-2148

GFT3108/80	AD132/II-142, ADY23-28, ADZ12, AUY28-32, • MP1531, NKT403, 2N514B-1531-2147
GFT3408/20	AD130/IV-142-149-150-153, ADY23, AUY33, MP1529, NKT405, 2N512A-1529-2148-2836
GFT3408/40	AD131/IV-138/50-142-149-150, ADY24-28, AUY32, MP1530, NKT405, OC26, TI3029, 2N1530-2065-2148-2836
GFT3408/60	AD131/IV-142-149-150, ADY24-28, AUY33, ASZ18, MP1530, NKT405, 2N512A-1530-2148
GFT3408/80	AD132-142, ADY23-28, ASZ15, AUY28-32, MP1531, NKT403, 2N513B-1531-2147
GFT3708/20-80	AD130-131-132-143-149-150-153, ADY15-23, ASZ15-18, AUY33, MP1529/30/31, NKT405-451, TI3029,.2N1529/30/31-2066-2148
GFT4012	OC26
GFT4012/30/60	AD130/III-142-149-150-153, ADY24, AUY33, MP1529, NKT405, OC26, TI3029, 2SB83, 2N1529-2063-2836
GFT4112/30-60	AD130-138-143-149-150-153, ADY24, AUY33, MP1529, NKT451, 2N513-1529-2148
GFT4308/40-80	AD130/1/2-142-149-150-153, ADY23-25, ADZ12, ASZ15, AUY23-25-28, MP1529/30/1, NKT405-451, TI3029, 2N257-513-1529-1530/1-2147/8
GFT4412/30-60	AD130/1-138-143-149-150-153, ADY24-28, ADZ12, AUY32, MP1530, NKT541, TI3029, 2N512/3-513-1529/30-2148
GFT4608/60	AD131/V-138/50, 142, ADY24-28, ADZ12, AUY32, MP1530, NKT403, TF3029, 2N513-1530-2066-2148
GFT4712/30-60	AD130-131-138-143-149-150-153, ADY23, AUY33, MP1529, NKT405, TI3029, 2N514-1529-2066-2148
GFT8024	AD149, CTP1514, OC26, 2N2836
GFY50	OC170
GI2711/2	BC170A-208/B-238A, MPS2711, SK3122, 40400, 2N3011-3298-5520
GI2713/4	BC208, BCY58, BSX51/2, MPS2713/4, SK3122, 2N3011-3394-5220, 40398
GI2715/6	BC172-208-238, MPS2715/6, 40232/3, SK3122, 2N3011-3394-5225
GI2921/2/3/4/5/6	BC168-170-208, BCY58, BSX51, MPS2921/2/3/4/5/6, TIS97/8, 40397/8, 2N2921/2/3/4/5/6-5225
GI3392	BC170-208-238, BCY58, MPS3392, TIS98, 40398, 2N3392-5220
GI3638	BC177-208-260, BCY79, BSW22-44, MPS3638, SK3114, TIS86, 40454, 2N3638-5225
GI3641	BFY34-67, BSY44-53, FT3641, ME6101, SK3122, 2N2224-3641-1613-5449
GI3642	BFY46-68, BSY54-71, FT3642, 2N1711-2221 3642-5449
GI3644	BFW20, BFX29, BSV16XVI, FT3644, MEO402, SK3114, 2N2905A-3503-3644-3962-4037-5447
GI3702/3	BC327-360, BSX36, ME0401, MPS3702/3, SK3114, 40319, 2N2904-3702/3-5447/8

GI3704 to 3711	BC282-289-337-340, ME8001, MPS3704, SK3122, 40084-40451/2/5, 2N2221-3704/11-5136-5449
GI3793/4	BFW64, BSX20, BSY19-21, ME9001, TI84, 40451, 2N2369-3566-3793/4
GM290	AF139, M3091, MM1139, 2N2244-5043
GM290A	AF139-239, BF272, AFY34-40, MM1139, 2N1142- 3127-3324-5244
GM378A	AF239, AFY16, BF272, MM1139, 2N1143-3127- 3323-5244
GM656A	AF239, AFY34-40, BF272, MM1139, 2N1142-3127- 3324-5244
GMO760	AF106-109-180-190, SK3114, 2SA432
GMO761	AF106-190, 2SA230
GS506/7	2N1306
GT3	AC125/6, 2N2429
GT4A	AC128-132, 2N2431
GT11/2	AF127-185, OC45, SK3005
GT13	AF126-185, OC44, SK3005
GT14	AC128, ASY77, OC72, SK3004, 2N2431
GT14H/20H	OC58-66
GT20	AC128-132, OC72, SK3004, 2N2431
GT31-34	AC125-126, SK3004, 2N2429
GT32/3	AC128-132, SK3004, 2N2431
GT34HV	ASY77, OC77
GT34N	ASY24-48-71-81, NKT217, BCW86, 2N1926-1998- 4030-5022
GT34S	AC125, SK3005
GT38	AC125-126, OC71, 2N2429
GT41/2/3	AF185, ASY27, SK3005
GT45	ASY26/7
GT70	ASY26
GT74-81	AC122-125/6-136-151-173, BC260A, BSX36, NKT219, SK3004, 2N1191/3-1352-1373-1384- 2429-2431-5040
GT81H	AC126, OC58-66, 2N2429
GT81R	AC128-132, SK3004, 2N2431
GT83-87	AC125-126, SK3005, 2N2429
GT88	ASY26, SK3005
GT109	AC122-125-128-136-151-173, BC260A, BSX36, NKT219, OC72, SK3004, 2N1193-1352-1375- 1384-2431-5040
GT109R	AC128-132, 2N2431
GT122	AC122-125/6-136-151-173, BC260A, BSX36, NKT219, OC76, SK3004, 2N1192-1371-1707- 2613-5040
GT123	ASY26, BSW72, BSX36, NKT135, SK3005, 2N1192- 1371-2613-2927
GT161	AF126-185
GT167	AC127-176-181, ASY28, BSX12, NKT717, SK3010, 2N1605A-2430-5135
GT222	AC121-125/6-173-178, BSX36, NKT281, OC71, SFT353, SK3004, 2N1191-1303-2429-2613-5040
GT229	AC127-130-176-181, ASY29, BSX12, NKT717, SK3011, 2N1605A-1808-2430-5135

GT310	AC128, 2N2431
GT758	AC122-125-136-151-173, BSX36, NKT219, SK3004, 2N1192-1371-1384-5040
GT759	AF127, OC45, SK3005
GT760/R	AF127-185, OC45, SK3005
GT761	AF126-127-185, SK3005
GT761R/62	AF126-185, OC44, SK3005
GT792	AC127-130-176-181, ASY28, BSX12, NKT717, SK3011, 2N1302-1605A-2430-5135
GT904	AC127-176-181, ASY29, BSX12, NKT713, SK3011, 2N1304-1605A-2430-5135
GT948	AC127-176-181, ASY29, BSX12, NKT717, SK3011, 2N1306-1605-2430-5135
GT949	AC125/6/7-130-176-181, ASY29, BSX12, NKT717, AF126, SK3010, 2N1308-1605A-2430-5135
GT1604	AC122-125-136-151-173, BSX36, NKT219, SK3005, 2N1190-1372-2613-5040
GT1605	ASY26, BSW72, BSX36, NKT210, SK3005, 2N1192-1373-1384-5040
GT1606/7	AF124-134-144, BFX48, BSW19-72, NKT674, SFT316, SK3005, 2N1524-2189-3323-4916
GT1608/9	AC127-130-176-181, AF167, BFX48, NKT674, SK3011, 2N1527-2191-3324-4916
GT5116/7	AF126-136-143, BFX48, BSW19-72, FT1746, NKT613, SFT316, SN3008, 2N1526-2273
GT5148/9	AF125-134-144, BFX48, BSW19-72, FT1746, SFT316, SK3006, 2N1525-2190-3324-5148-5354
GT5151/3	AC117-124-128-142-153-180, BCW86, NKT229, SK3006, 2N1926-1998-4030-5022
H2/3/4	AD149, 2N2836
H8DEF	AF185
HA1/2/3	AC126, OC66-71, 2N2429
HA8/9/10	OC66
HCl	AC126, 2N2429
HD197	AC128, 2N2431
HF1/2	AF185, SK3018
HJ15	SK3004, 2N362-2429
HJ17	AC128, SK3004, 2N2431
HJ17D	AC124, SK3004, 2N632-2431
HJ22	AF185, SK3005
HJ22D	AF131-137-185, SK3005
HJ23	AF126-185, SK3005
HJ23D	AF185, SK3005
HJ32	AF135
HJ34	2N2431
HJ34A	SK3004, 2N2431
HJ50	SK3004, 2N2429
HJ51	AC180, SK3004, 2N2431
HJ54	SK3005
HJ55/57	AF132-136-185
HJ56	AF133-137-185, SK3008
HJ60	AF132-136-185, SK3005
HJ61	AF133-137

HJ62	AF185, SK3005
HJ70	AF132-135-185, SK3007
HJ71	AF185, SK3005
HJ72	AF185
HJ73	AF185, SK3005
HJ74	AF132-137, SK3005
HJ75	AF185
HS3/4	ASY27, OC47
HS15/17D/22D	SK3004
HS23D	SK3004
IF1/2/3	AF185
IMF3954/A-5/A	E413
IMF3956	E414
IMF3957/8	E415
IT108	E304
IT200	E211
IT210	E210
IT220	E305
ITE3066/4119	E203
ITE3067/4118	E202
ITE3068/4117	E201
ITE4338/9	E201
ITE4340	E202
ITE4341	E203
ITE4391	E111
ITE4392	E112
ITE4393	E113
ITE4416	E304
ITE4867/8	E230
ITE4869	E231
IW8377	BF155-180, BFX62
J1/2/3	OC72
JP1	AC128, OC72, 2N2431
KC147	BC147, ZTX107
KC148	BC148, ZTX108
KC149	BC149, ZTX109
KC507	BC107
KC508	BC108
KC509	BC109
KCY58	BCY58
KCY59	BCY59
KCZ58	BFY91, 2N2915
KCZ59	BFY92, 2N2917
KD601	BD109
KF124	BF194
KF125	BF195
KF167	BF167
KF173	BF173
KF503	BF114-177
KF504	BF110-117-178
KF506	BFY34, 2N1613
KF507	BFY33
KF508	BFY46, 2N1711
KF517	2N1131/2-1991

KF521	3SK21
KF524	BF184
KF525	BF185
KF2000/2	ACY18-24, ASY48, BCW86, NKT217, 2N1926-2890-5022
KF2001/3	ACY39, BCW86, 2N398-2042-2890-5151
KFY16	BFY16
KFY34	BFY34
KFY46	BFY46
KJ2001/3	ADY24-28, ADZ12, AUY22-28-32, MP1551, NKT401, 2N1021-1906-3615
KJ2002/4	ADY25-28, ADZ12, AUY28-32-34, MP1552, NKT403, 2N1022-1906-3610
KM7000/7	AD149-150-153, ADY23, ASZ16, NKT406, TI3028, 40051, 2N2143A-2553-2556
KM7008	AD134-138/50-153, ADY23-28, ADZ11, NKT406, 2N1530-2138-2557-2869
KM7009	AD135-138/50-153, ADY24-28, ADZ12, NKT401, 2N1531-1906-2139-2558
KM7010	AD163, ADY25, ASZ18, AUY28, NKT420, 2N1532-1906-2140-2559
KM7011	AD133-138, ADY23, ASZ16, NKT404, 40051, 2N1146-1544-2143
KM7012	AD134-138/50, ADZ11, ADY16, ASZ16, 2N1146A-1545-2144-2869
KM7013/6	AD135, ADY28, ADZ12, AUY28-30, NKT403, 2N1146B-1546-2145-2870
KM7014/7	AD163, ADY25, ASZ18, AUY28, NKT420, 2N1146C-1547-2146-2870
KM7015	AD134-138/50, ADY23, ADZ11, ASZ15, NKT402, 2N1147A-1540-1548-1906
KR206	BR100
KS500	BSY62
KSY04	2N2904
KSY21	BSY21, 2N914
KSY34	BSY34
KSY62	BSY62
KSY63	BSY63
KSY71	2N2369
KSY81	2N2894
KU601/11	BDY12B/VI
KU605/7/12	BUY12
KU606	BUY13
LDA400	BC208, BCY58/VI, BF237, ME2001, MPS6575, NKT10339, SE2001, 40244, 2N929
LDA401/2/3	BC208, BCY58A, BF237, ME2002, MPS6565, NKT10439, SE2002, 40245, 2N930
LDA404	BFY34-67, BSY44-53, EN697, ME6001, SK3024, 2N697-1613-2538
LDA405	BFX68, BFY46-68, BSY54-71, ME6002, EN956, SK3024, 2N956-1711-2537
LDA406	BSV89, BSY19-63, ME3001, MM1501, NKT16229, SE3001, SK3024, 2N916-2369-2865-3600

LDA407	BF357, BFX59, BFY39-88-95A, BSV90, ME3002, MM1500, SE3002, 2N744-2839
LDA410	BSX20-72, BSY63-87, BSV92, BF224, ME9001, MM5001, NKT10439, SE5001, 2N2368-2839
LDA450/1/2/3	BC291D-327-360VI, BSW21A, ME0404, MPS3703, NKT20339, SE8540, SK3025, 40319, 2N3703-5448
LDF603	E305
LDF604	E304
LDF605	E210
LDF691	E111
LDF692	E112
LDF693	E113
LDS200	BFX59, BFY39-88, BSV89, BSY95A, BF224, ME4001, MM8000, NKT16229, SE4001, 2N2368-9
LDS201/5/7	BF224, BSV89, BSY19-63, MM8001, ME4002, NKT16229, SE4002, SK3024, 2N2369
M10H	AF178, 2N2495
M10L	AC176
M12H	AF178, 2N2495
MA100	ACY24, ASY48-77, BCW86, SK3051, 2N398-2890-5022
MA200/1/ 2/3/4/5	ACY39, BCW86, 2N398-2980-5151, SK3016, NKT217
MA206/881/ 882/883	ACY24, ASY48-77, BCW86, NKT217, 2N398-2890-5022
MA884/5/ 6/7/8/9	ACY24, ASY48-77, ME0402, NKT217, 2N398-2890-4031-5022
MA909/10	ACY24, ASY48-77, BCW86, ME0401, NKT217, 2N398-2890-4030-5022
MA1703/4/ 6/7	AC117-128-142-153-180, BFS12, BSW72, NKT211, SK3004, 2N1998-2000-2376-5041
MC101	AF185, SK3007
MD420	AC120-128-142-178-180, BFS12, BSW72, NKT211, SFT323, 2N1999-2000-2376-5040
ME213/A	BC207, BCY59A, MPS6553, NKT10439, SK3122, 2N3392-5135-5449, 40398
ME216/7	BC208, BCY58A, MPS6553, NKT12329, SK3122, 40231
ME501	BC192-328, BSX12, SK3114, 2N996-2894-5354
ME502/3	BC192-328, BSX36, SE8540, 2N3121-4402-5023-5354/5
ME513	BCW86, BFX30-41, BSV16VI, BSW74, FT4356, 2N4031-4236-4356-5365
ME900	BFY39, BSX20, BSY21-63, NKT10339, SK3122, 2N914-2369-3137-3227
ME901	BFY39, BSW41-83, BSX25-48, NKT10429, SK3122, 2N2369A-3227-4137-4427
ME1001/2	BC182A, BFY39-55, BSX45-51A, MPS6575, NKT10339, SE1001, SK3122, 40244, 2N929
ME1075	BF156-177, BSS34, SK3024, 40360, 2N297-1990-3114
ME1100	BF118-156-178-186-256-297, MM3001, SE7002, 40349, 2N3114

ME1120	BF117-156-177-297, MM7002, SE7002, SK3045, 40349, 2N3114
ME2001	BFX62, BFY39I, BSX38-51A-69-81, MPS657, NKT10339, SE2001, SK3122, 40244, 2N929-3365
ME2002	BFX59, BFY39II, BSX52A-69-81, MPS6575, NKT10439, SE2002, SK3018, 40245, 2N930-3565
ME3001	BF357, BFX59, BFY39-66, BSX95, BSY95A, MM1501, NKT16221, SE3001, SK3018, TIS62 40480, 2N2369-3011
ME3002	BF357K, BFX59, BFY39-66, BSX93, BSY95A, MM1501, NKT16229, SE3002, SK3039, TIS63, 40482; 2N2369-3011
ME3011	BFX59, BFY39-66, BSX93, BSY95A, MM1501, NKT16229, SE3005, SK3039, SX18, TIS64, 40482, 2N2369-3011
ME4001/2/3	BC168A-184-208, BCY58A, MPS6574, NKT10419, SE4001, SX3710, 40398, SK3020, 2N3706
ME4101/2	BC167A-207/8-385, BCY59A, ME4001, MPS3705, SE4001, SK3122, 40472, 2N3705
ME4103	BC167C-182-207-386, BCY59C, MPS3706, SE4003, SK3122, 40237, 2N3706
ME4104	BC182A, SK3122
ME5001	BF115, BFX55, BFY88, BSV92, GI3693, MPS3693, NKT16229, SE5001, SK3039, 2N2708-3693-3866
ME6001	BSV92, BSX48-71-72, BSY81, NKT10439, SE6001, SK3122, 40405, 2N2368-3706-5448
ME6002/3	BSV89, BSX48-71-75, BSY82, MM8001, NKT10439, SE6002, SK3122, 40519, 2N2369-3706-5447
ME6101	BFR41, BFX67, BFY34-67, BSY44-53, EN697, 2N697-2410
ME6102	BFR41, BFY46-68, BSY54-71, EN956, 2N956-1711-3109-3569
ME8001	BC211, BFR51, BSV89, BSX49-71-75-82, SE8040, SK3024, 2N697-2410-2501-2787-5450
ME8002	BFR21-39, BFY14B, BSW10, BSV64, BSY87, SE8010, SK3024, 2N911-1893-1974
ME8003	BFR40, BFY46-68, BSY54-71, SE8012, SK3024, 2N1711-3569
ME8101	BFY39, BSX72, BSY19-95A, BSV90, EN744, 2N744-4264-5187
ME8201	BFX12, BSX29, BSW19-21, SK3039, TIS54, 2N2894/A-2576-5055-5292
ME9001	BFW68, BSW63, BSX20, BSY21, EN916, SK3039, TIS48, 2N916-2432-3211-3261-4954
ME9002	BSX20-92, BSY19-63, EN3013, SK3018, TIS51, 2N703-3011-3261-3645
ME9021	BSW92, BSX20, BSY19-63, EN3009, SK3039, TIS48, 2N3009-3201-3046
ME9022	BSW82, BSV89, BSY19-63-70, MM8001, SE6003, SK3039, TIS51, 2N2369-3261-3706
ME0401	BC291D-360VI, BFR81, BFX29, BSV16VI, FT3644, 40362, 2N2906-3250-3485-4142

ME0402	BC291A-360VI, BFR81, BFX30, BSV16VI, FT3645, 40362, 2N2907-3251-3486-4143
ME0404	BC328-360VI, BSV16VI, BSX39, GI3702, MPS3702, NKT20329, SK3025, 2N2904-5447-5451
ME0404-I/II	BC327-361VI, MPS3703, NKT20339, SK3114, 2N2904-3703-5448-5450
ME0411/2	BC177A-212A-307VI-292A-297, BSV16X, SK3114, 2N3798-3962-4354-5366
ME0413	BC213-381, BFX29, BSV15VI, BSW72, BSX36, NKT20339, SK3114, 40406, 2N3307-5365
ME0461/2	BC212-219D, BFX29, BSV16VI, BSW74, FT3644, 40362, 2N2904-3250-3485-4124
ME0463	BC328, BSV15VI, BSW72, BSX36, MPS3703, NKT20339, 40319, 2N3703-5366-5477
ME0475	BC328, BFR79, BSV15VI, BSW72, BSX36, MPS3702, NKT20329, 40319, SK3118, 2N3702-5366-5448
ME0491	BFX52, BSW44A-81, BSY40, FT4356, TIS54, 2N4031-4236-5448
ME0492	BFS53, BSW44-81, BSY41, MPS3703, NKT20339, 40319, 2N3703-5448
ME0493	BSW81, BSX29, BSY41, NKT20339, TIS50-54, 40319, 2N2894-5448
MFE2000	E305
MFE2001	E304
MFE2004/5/7	E112
MFE2006/8/10	E111
MFE2009/11	E110
MFE2012	E108
MFE2093	E201
MFE2094	E201
MFE2095	E203
MFE2133	E112
MFE3001	E202
MFE4007/8	E202
MFE4009	E305
MFE4010/1/2	E305
MFT122	AC128, 2N2431
MFT123	AC128, 2N2431
MHT4401/ 2/3/4	2N4300
MHT5505/ 6/7/8	2N4300
MHT6408 to 6416	2N3996
MHT7011 to 7019	2N4301
MHT7401 to 7419	2N3421
MHT7801 to 7809	2N5387
MHT8002/4	2N4002
MHT9001 to 9012	2N4002

MJ423/31	BDY96
MJ450	2N4398
MJ480	SK3027, 2N4913
MJ481	SK3027, 2N4914
MJ490	2N4904
MJ491	2N4905
MJ1800	BDY98
MJ2255/6	2N3713
MJ2257	2N3714
MJ2267	2N4901
MJ2268	2N4902
MJ2801	SK3027, 2N3713
MJ2802	SK3021, 2N3714
MJ2901	2N4901
MJ3029	BDY95
MJE101	BD132, TIP34
MJE102/4	TIP34A
MJE103	TIP34
MJE105	BD132, TIP34A
MJE201	BD131, TIP33
MJE202/4	TIP33A
MJE203	TIP33
MJE205	BD131, TIP33
MJE340	BD144, SK3021
MJE370	SK3025, TIP32
MJE371	SK3083, TIP32
MJE520	SK3041, TIP31
MJE521	SK3054, TIP31
MJE2801	TIP33A
MJE2901	TIP34
MJE2955	TIP34
MJE3055	TIP33A
MK10	E305, SK3116
MM380	AF239, AFY16, BF272, 2N1141-3307-3324-5056
MM709	BFY37-39, BSV90, BSX72, BSY17, ME8101, EN709, SK3039, 2N744-4268-5187
MM1139	AF239, AFY16, BF272, 2N1142-3307-3324-5057
MM1755/7	BFX67, BFY34-67, FT3642, ME6001, SK3122, 2N697-2538-3642
MM1756/8	BFX68, BFY46-68, BSY54-71, FT3643, ME6002, SK3122, 2N956-2537-3643
MM1803	BFY33-67C-72-81, BSX79, BSY51, ME1075, SK3039, 2N696-1984-2410
MM1941	BF224, BFX59, BFY39-66, BSV89, BSY95A, ME9002, NKT16229, SE1010, SK3039, 2N2857
MM2483	BFX67, BFY34-67, BSY44-53, ME6001, 2N929-2483-2791
MM2484	BFX68, BFY46-68, BSY54-71, ME6002, 2N930-2484-2792
MM5000	AF239S, AFY39-40, BF272, ME0191, 2N1141-3307-3324-5057
MM5001/2	AF239S, AFY18-42, BF272, 2N1141-3307-3323-5057

MM8000/1/2	BF224, BFY39-88, BFX59, BSV89, BSY95A, ME4001, NKT16229, SE4001, 2N2369-2839
MMF1/2/3/4/5/6	E400
MMR6/1	AC126, 2N2429
MMR6/2	AF178, 2N2495
MMR6/3	AF185
MMR6/11	AC128, 2N2431
MMR6/12	AD149, 2N2836
MN24/5/6	AD149, OC16, SK3009, 2N2836
MP939	AU103-105
MP1529	ADY26, AUY21-28, DTG600, SFT265, 2N511-1031-1359-1529
MP1530	ADY26, AUY22-28, DTG601, SFT266, 2N375-457A-511A-1031A-1530
MP1531	ADY26, AUY28-34, DTG602, SFT267, 2N375-511B-638-1031B-1531
MP1532/3	AUY28-34, DTG603, SFT268, 2N630-638A-1031C-1100-1362-1532-1906
MP1534	ADY26, AUY21-28, DTG600, SFT265, 2N677-1032-1534-1539
MP1535	ADY26, AUY22-28, DTG601, SFT266, 2N457A-677A-1032A-1535-1540
MP1536	ADY26, AUY28-34, DTG602, SFT267, 2N458A-677B-1032B
MP1537	ADY26, AUY28-34, DTG603, SFT268, 2N637-677C-1032C-1537
MP1538	AUY28, DTG603, SFT268, 2N673B-1365-1100-1538
MP1549	ADY26, AUY21-28, DTG600, SFT265, 2N513-678-1162-1539-1549
MP1550	ADY26, AUY22-28, DTG601, SFT266, 2N513A-678A-1164-1540-1550
MP1551	ADY26, AUY28-34, DTG602, SFT267, 2N513B-678B-1165-1541-1551
MP1552	ADY26, AUY28-34, DTG603, SFT268, 2N678C-1100-1166-1542-1552
MP1553	ADY26, AUY21-28, DTG600, SFT265, 2N277-677-1031-1543-1553
MP1554	ADY26, AUY22-28, STG601, SFT266, 2N278-677A-1031A-1544-1554
MP1555	ADY26, AUY28-34, DTG602, SFT267, 2N677B-1031B-1099-1545-1555
MP1557	ADY26, AUY21-28, DTG600, SFT265, 2N1032-1146-1547-1557-4278
MP1558	ADY26, AUY22-28, DTG601, SFT266, 2N1032A-1146A-1548-1558-4279
MP1559	ADY26, AUY28-34, DTG602, SFT267, 2N1032B-1146B-1549-1559-4281
MP1560	ADY26, AUY28-34, DTG603, MP2300, SFT268, 2N908-1032C-1146C-1560
MP1612/A/B	AU101-105, DTG2200, SDT1860, 2N908-1032C-1146C-1560
MP1613	ADY23-26, AUY28-30-34, SFT268, 2N678B-1166-1542-1552

MP2060/1	ADY24-26, AUY21-28-31, SFT265, SK3009, 2N678-1162-1539-1549/50
MP2062/3	ADY25-26, AUY22-28-30/1, SFT267/8, 2N678A/B-1165/6-1541/2-1551/2
MP4104	E304
MPF102	E305, SK3116
MPF103	E305, SK3112
MPF105	E304, SK3112
MPF106/7	E304, SK3112
MPF108	E305
MPF109	E202, SK3112
MPF111	E202
MPF161	E174
MPS292	BC238B
MPS370	BC238
MPS404/A	SK3114, 2N4059
MPS653	BC237A
MPS706/A	SK3039, TIS44
MPS708	TIS45
MPS834	SK3039, TIS52
MPS918	SK3039, SX18, TIS62
MPS2369	SK3039, TIS48
MPS2711/2	BC170A-208-238A-386, GI2711, NKT10419, SK3122, 40400, 2N3709-5220
MPS2713	BC208-386, BCY58A, BSX51, GI2713, NKT10419, SK3122, TIS47, 40398, 2N5220
MPS2714	BC208-386, BCY59B, BSX52, GI2714, NKT10519, SK3122, TIS48, 40398, 2N5220
MPS2715	BC172A-208-238A-386, GI2715, NKT10519, SK3122, 40232, 2N5225
MPS2716	BC172B-208-238B-386, GI2716, NKT10519, SK3020, 40233, 2N5225
MPS2921/2	BC208-386, BCY58A, BSX51, GI2921/2, NKT10519, 40398, 2N5225
MPS2923	BC168A-170A-208, BCY58A, BSX51, GI2923, NKT10419, SK3122, TIS99, 40398, 2N3710-5220
MPS2924	BC168A-172A-208-385, BCY58A, BSX51, GI2924, NKT10419, SK3122, TIS98, 40397, 2N3711-5225
MPS2925	BC168B-170B-208, BSX52, GI2925, NKT10519, SK3122, TIS98, 2N3711-5220
MPS2926	BC168C-170C-208-385, BSX52, GI2926, NKT10519 SK3122, TI2926, 40397, 2N5220
MPS3392	BC168B/C-172B/C-208-209C, BSX52, GI3392/3, NKT10519, SK3020-3122, 40397, 2N3710-3707/8-5136
MPS3394	BC168A-172A-208, BSX51, GI3394, NKT10419, SK3122, 40398, 2N3708/9
MPS3395	BC168C-172C-208, BSX52, GI3395, NKT10519, SK3122, 40397, 2N3711-5136
MPS3396	BC100D-172D-208, BSX51, GI3396, NKT10419, SK3122, 40398, 2N3708-5135
MPS3397/8	BC168-172-208, BCY58, GI3397, NKT10519, SK3122, 40398, 2N3708

MPS3563	SK3039, TIS62
MPS3638/A/40	SK3114, TIS50
MPS3646	SK3046, TIS55
MPS3693/4	BF594J, SK3018
MPS3702	BC192-328, BFW20, ME0401, NKT20329, SK3114, 2N3702-5365-5447
MPS3703	BC327, BFW23, BSW74, ME0401, NKT20329, SK3114, 40406, 2N3703-5366-5448
MPS3704	BC337-341VI, BFX17, ME6101, NKT10339, SK3122, 40084, 2N3702-3704-4954-5448
MPS3705	BC327-341VI, BFX17, ME6101, NKT10329, SE3705, SK3122, 40451, 2N3705-5450
MPS3706	BC337-341X, BFX17, ME6101, NKT10439, SE3705, SK3122, 40451, 2N3706-5451
MPS3707/8	BC338-340VI/X, BFY61, ME6101/2, NKT10439-10519, SK3122, 40452, 2N3707/8-4951/2
MPS3709/10	BC208-338-340X, ME6102, NKT10519, SK3122, 40455, 2N3709/10-4952/4-5136
MPS3711	BC208-338-340XVI, ME6102, NKT10519, SK31220, 40456, 2N3711-4954-5136
MPS3712	BC209-239, BCY59, GI2921, NKT10519, 40398, 2N3708-5136-5449
MPS3721	SK3122
MPS5172	2N3710
MPS6506	TIS86, 2N4996
MPS6507	BF224J, SK3018, TIS86, 2N4996
MPS6508	TIS86, 2N4996
MPS6509/10	TIS84
MPS6511	BF224J, BFW63, BFY66, BSY19-63, ME2001, NKT12429, SK3039, TIS56;87, 2N2369-3013-3932
MPS6512/3	BC207-237, BCY59A, BFX17, GI3704, ME6001, SE6001, NKT10339, 40084, SK3122, TIS99, 2N3709-5459
MPS6514	BC207-237B, BCY59B, BFX17, GI3706, ME6002, NKT10439, SE6002, SK3018, TIS98, 40451, 2N3711-5450
MPS6515	BC184-207-237C, BCY59C, BFX17, GI3708, ME6002, NKT10439, SE6002, SK3122, TIS97, 40452, 2N5450
MPS6516/7/8	BC212, SK3114, 2N5447
MPS6519	BC214, SK3114, 2N5447
MPS6520/1	BC184-207-237-386, BCY59C, BFX17, GI3706, NKT10439, SE6002, SK3122, 40451, TIS98
MPS6522/3	BC214, SK3114, 2N5447
MPS6528	SK3039, TIS84
MPS6529	SK3039, TIS108
MPS6530	BFR39, SK3024, TIS92
MPS6531	BFR41, SK3020, TIS92
MPS6532	BC237A, BFR50, SK3039, TIS92
MPS6533	BC177IV, BFR79, SK3114, TIS93
MPS6534	BC177A, BFR81, SK3114, TIS93
MPS6535	BC177IV, BFR60, SK3114, TIS93

MPS6539	BF224, BFX59, BSV92, BSX20, BSY19, EN3013, ME9002, NKT10229, SK3039, 2N703-3014-4427-4875
MPS6540	BF224, BFX59, BSV90, BSX20, BFY19, ME9002, NKT16229, SK3018, 2N2369-3014-3566-4427-4876
MPS6541	SK3039, TIS86, 2N4996
MPS6542	BF224, SK3039, TIS62
MPS6543	BF224, BFX59, BSV92, BSX20, BSY19, GI3793, ME9001, NKT16229, SK3019, TIS86, 2N2369-3839-5130
MPS6544	BF224, SK3018, TIS87
MPS6545	BF224, SK3122, TIS87
MPS6546/7/8	BF224, BFX59, BSV89, BSY19, GI3794, ME9001, NKT16239, SK3039, TIS86, 2N2369-3839-5130
MPS6551/2/3	BC222-338, BSY81, GI3706, ME6001, SE8040, 40311, 2N3416-3706-5451, TIS98, 40311
MPS6554/5	BC282-338, BSY82, GI3706, ME6002, SE8040, SK3122, 40311, 2N3417-3706-5451
MPS6560/1	BC338-340VI, BFR51, BSW42, SE8040, SK3122, 40311, TIS90-92, 2N696/7
MPS6562	BFR61, SK3114, TIS93
MPS6563	BC328-360VI/X, BFR62, BSW44/5, SE8540, SK3114, 40406, 2N1131/2, TIS91/3
MPS6564	2N4994
MPS6565	SK3122, TIS99, 2N3709
MPS6566	SK3020, TIS98, 2N3710
MPS6567	BC208, BCY58, BF357-594, ME6001, SE6001, SK3122, 40311, TIS86, 2N3565
MPS6568/A	BF225
MPS6569/70	BF225, SK3039
MPS6571	BC184
MPS6579	TIS37
MPS-A10/20	2N3708
MPS-A12/14	2N5526
MPS-A13	2N5525
MPS-A55	BFR80
MPS-A65/6	2xBC214
MPS-A70	2N4059
MPS-A05	BFR40
MPS-A06	BFR39
MPS-A09	2N3707
MPS-H10/1	BF224
MPS-H20/30/31/32	BF225
MPS-H24/34	BF224
MPS-H37	BF594
MPS-H54/5	BFR79
MPC-H03	BF225
MPS-H04/5	BFR39
MPS-H07/8	BF225
MPS-L51	BF398
MPS-L01	BF297

MPS-L07	TIS53
MPS-L08	TIS54
MPSU05	BD137
MPSU06	BD139
MTC70/2/6	AC128, 2N2431
MTC71	AC126, 2N2429
MU4891	2N4891
MU4892	2N4892
MU4893	2N4893
NF506/50	E305
NF4302/4	E202
NF4303	E203
NKT11	AC122-125-136-173, BC260A, BSX36, 2N1191-1305-1352-1371-1384-5040
NKT12	AC122-125-136-151-173, BC260A, BSX36, 2N1191-1306-1352-1373-1384-5040
NKT52	NKT72
NKT53	NKT73
NKT54	NKT73
NKT62	NKT72, SK3005
NKT63/4	NKT73, SK3005
NKT72	AC122-125-136-151-173, BC260A, BSX36, SK3005, 2N1192-1352-1373-1384-5040
NKT73	AC117-131-132-142-152-180, BC126-160A, SK3005, 2N1189-1925-1998-2375-5022
NKT74	NKT73, SK3005
NKT121	2N1309
NKT122/3	2GT102, 2N1305
NKT124	AC117-131-132-142-152-180, BC126-260A, 2N1309-1925-1998-2376-5022
NKT125	AC122-125-136-151-173, BC260A, BSX36, 2GT102, 2N1191-1305-1352-1371-1384-5040
NKT126/9	2GT102, 2N1303
NKT127	SK3006, 2N1309
NKT128	2GT102, 2N1305
NKT132	AF185, SK3005
NKT133	AF185, SK3004
NKT135	AC117-131-132-142-152-180, BC126-260A, 2N404A-1925-1998-2375-5022
NKT137	AC117-131/2-142-152-180, BC126-260A, 2N1925-1998-2375-5022
NKT141	SK3005
NKT142/3	2GT102, SK3005
NKT144	SK3005, 2N1303
NKT152	NKT162, SK3006
NKT152C	NKT162
NKT153/A/B	NKT163
NKT153/25	NKT163, SK3006
NKT153/25B-C	NKT163
NKT154/A/B/R	NKT164
NKT154/25	NKT164, SK3006
NKT154/25B/C	NKT164
NKT162	2GT102, SK3005

NKT163/25	NKT163, SK3005
NKT163/25B	NKT163
NKT164/25	NKT164, SK3005
NKT172	NKT162
NKT173	2G371, NKT163
NKT174	NKT164
NKT182	NKT162
NKT183	NKT163
NKT184	NKT164
NKT202	SK3005, 2N2429
NKT203	AC126, SK3004, 2N2429
NKT204	SK3004, 2N2429
NKT205	SK3004, 2N2429
NKT206	SK3005, 2N2429
NKT208	2N2431
NKT210	AC117-131-132-142-152-180, BC126-260A, 2N1926-1999-2376-5022
NKT211	AC117-124-142-153-184, BC126-260A, 2CY18, 2N1998-2000-2374-3427-5022
NKT212	AC117-142-152-180, BC126-260A, 2G303, 2N1998-2000-2375-3427-5022
NKT213/4/5	AC117-131/30-132-152-180-192, BC126-260A, 2G374, 2N1925-1998-2000-2375-5022
NKT216	AC117-132-152-180-192, BC126-260B, 2G374, 2N1998-2000-2374-3427-5022
NKT217	AC192, ACY24, ASY48-77-81, BCW86, 2G377, 2N1926-1999-2000-2375-4030-5022
NKT218	2CY21
NKT219	AC117-131/30-152-190, ASY80, BC126-260A, SFT323, 2N1305-1998-2000-2376-3427-5022
NKT221	2CY20, SK3005
NKT222	2CY21, 2GT182, SK3004
NKT223	AC117-131/30-152-156-192, BC126-260A, 2CY21, SK3004, 2N2000-2376-2614-3427-5022
NKT224	AC131/30-152-192, ASY80, BC126-260A, SK3004, 2N2000-2376-2614-2800-5022
NKT225	AC117-131/30-152-192, ASY80, BC126-260A, SK3004, 2N1303-527-2000-2375-2614-2800-5022
NKT226	SK3004, 2N1305
NKT228	SK3004, 2CY20
NKT229	AC117-131/30-152-192, ASY80, BC126-260A, 2N2000-2374-2614-2801-5022
NKT238/9	2N2000
NKT250	NKT274
NKT251	NKT271, 2N2431
NKT251A	NKT271
NKT252	AC126, NKT274, 2N2429
NKT252A	NKT274
NKT253	NKT271
NKT254	NKT274, SK3006
NKT254A	NKT274
NKT255	NKT275, SK3006
NKT255A/E/T	NKT275

NKT258	NKT274
NKT260	NKT274
NKT261	AC117-152-178-192, ACY38, BC126-260A, NKT271, SK3005, 2N1925-2303-2374-2614-3427
NKT261A	NKT271
NKT262	AC116-117-152, ACY38, BC126-260B, NKT274, SK3005, 2N1926-2303-2614-2837-3133
NKT262A	NKT274
NKT263	NKT271, SK3005
NKT264	AC116-117-162-192, ACY38, BC126-260A, NKT274, SK3005, 2N1926-1998-2303-2837
NKT265	NKT275, SK3006
NKT265A/E/J	NKT275
NKT268	NKT274
NKT270	AC117-152-192, ASY27, BC126-260A, NKT274, SFT351, SK3006, 2N1926-1999-2303-2838
NKT271	AC117-152-154-178-192, ASY78, BC126, BSW74, SFT321, SK3004, 26302, 2N1925-1998-2837/8-5022
NKT271A	NKT271
NKT272	AC122/3-113-152-173-192, ASY26, BC126-260A, NKT274, SK3004, 2N1926-1998-2837/8-5022
NKT272A	NKT274
NKT273	NKT271
NKT274	AC116-152-173-192, ASY26, BC126, BSW72, 2N2000-2374-2614-2801-5022
NKT275	AC116-122-152-173-192, BC126, BSW72, 2N2000-2375-2614-2801-5022
NKT277	NKT271
NKT278	NKT274
NKT281	AC122-125-151-191, BC126-260A, SFT351, 2N1191-1352-1371-1384-5040
NKT302	ACY24-38, ASY48-77-81, BCW86, 2N398-1998-2837-4030-5151
NKT304	AC117-122/30-152-192, ASY27, BC126-260A, 2N526-1198-2000-2801-2837-5022
NKT351	AC117-122/30-152-192, ASY27, BC126-260A, 2N1998-2000-2801-2837-5022
NKT401	ADY25-26, AUY34, MP2063, SFT268, SK3009, TI3030, 2N1100-1542-1552-1906
NKT402	ADY33, ADZ11, AUY21-28-30, MP2061, SFT266, SK3009, 2N1907-2869-4279
NKT403	AD140, ADZ12, AUY22-28-30, MP2062, SFT267, SK3009, 2N1541-2870-4281
NKT404/5/6	ADY23, ADZ11, AUY21-28-30, MP2062, SFT 266, SK3009, 2N457A-1540-2869-4279
NKT420	ADY25, ADZ12, AUY28-32-34, SFT268, 2N1022-1073B-1360-2869
NKT450	NKT452, SK3009
NKT451	ADY23, ADZ11, AD138, AUY21, SFT265, SK3009, 2N1147-1359-1549
NKT452/3	AD138-153, ADY23, ADZ11, AUY21, SFT265, SK3009, 2N278-1534-1539/A

NKT603F	AF121-193-202, BFX48, BSW72, GM1213B, 2N2273-3324-4916
NKT613F	AF121-193-202, BFX48, BSW72, 2N2273-3329-4916
NKT674F	AF124-134-170, BFX48, BSW19-72, FT1746, SFT358, 2N2273-2873
NKT677F	AF127-136-172, BFX48, BSW19-72, FT1746, GM1213B, SFT358, 2N2411-2635-2873
NKT713	AC157-179-181, BSX12, BSW82, SK3010, 2N1808-2430-5135
NKT717	AC127-179-181, BSW82, BSX12, 2N1605A-1808-2430-5135
NKT734	AC127, BSW82, BSX12, SK3011, 2N1605A-1808-2430-5135
NKT736	AC127, ASY28, BSW82, BSX12, SK3011, 2N1605A-1808-2430-5135
NKT751	NKT773
NKT773/81	AC127-179-181, ASY29, BSW82, BSX12, 2N1605A-1808-2430-5135
NKT10339	BC337, BSW28-84, ME6001, MPS6573, SE5020, SK3122, 2N3691-4874-4954-5183
NKT10419	BC168-172-208-235, BSW28, MPS6573, SE5021, SK3122, 2N3692-4874-5183
NKT10439	BC337, BSW85, BSX39, ME6002, SK3122, 2N834-3566-4123-4876-4954-5109
NKT10519	BC168-172-208, BSX39, SK3122, 2N834-3566-4123-4876-5109
NKT12429	BC338, BSW83, BSX39, ME8001, SK3122, 2N2330-3013-3566-3692-4427-4954
NKT13329	BFY72, BSX19-48, BSY21, ME6001, MPS6511, SK3122, 2N2368-3691-3932-3013
NKT13429	BFY72, BSY21, DSX20-48, ME6002, MPS6512, SK3122, 2N2369-3013-3692-3933-5449
NKT16229	BFY19-90, BSV90, BSY63-95A, ME8101, SE1010, 2N918-3011-352 3-3570-3600
NKT20329	BC328-338, BFX48, BSW72, ME404I, SK3114, 2N3644-4037-4059-4125-4916-5354
NKT20339	BC327-337, BSW73, BFX48, ME0404II, SK3114, 2N3644-4037-4059-4126-4917-5365
NKT35219	BFY19-78, BFW30, BSY62-95A, GI2711, ME8101 SK3117, 2N914-918-4252-5127-5179
NKT80111	E202
NKT80112/3	E203
NKT80211/2/3/4/6	E232
NKT80215	E231
NN7000/1/3/4	BFR40
NN7002/5	BFR41
NN7500/1/3/4	BFR80
NN7502/5	BFR81
NPO100/A	E304
NCP211/N	E201
NPC212N/4N/5N	E201
NPC213N	E202
OC13	AC125, OC71, 2N2429

OC14	OC4-72
OC16	AD130II-162, SK3009, TI3029, 2N1038-1314-2063-2836
OC20	SK3009, TI3030
OC22	AD148IV-149, AUY19V, SK3009, SFT213, 2SB338, 2N157A-301A
OC23/4	AD148V-149 AUY19V, GFT4012, SK3009, 2SB 338, 2N157A-1022-2869-3615
OC25	SK3009, TI3027, 2N2836
OC26	AD148-150/IV-153, SK3009, TF80, TI156-3027, 2N456-1022-1314-2836-2869-3615
OC27	SK3009, V30/30P, 2N353-1315
OC28	AUY22II-28, ASZ18, SK3009, TI3030, 2SB341, 2N174-1073-2870
OC29	ASZ18, AUY21III-31, AD138/50, SK3009, TI3028, 2SB341 2N174-1073-2870
OC30	AD148-148V-149-152-155-156V, SK3009, TF77-78/30III, TI3027
OC30A	AD148IV-148V-152-162, SK3009
OC30B	AD132III, SK3082
OC32	OC65
OC33	AC122-126-138-151-173, BC126-260A, NKT211, OC65, 2N1191-1352-1371-1384-2424-5022
OC34	AC122-138-151-173, BC126-260A, NKT211, SK3004, 2N1191-1352-1371-2613-5022
OC35	AD138/50, ASZ18, AUY21III-31, SFT214, SK3009, TI3028, 2SB339, 2N1073-2870
OC36	AUY22II-28, SK3009, TI3029, 2SB341, 2N157A-174-1073-2870
OC38	2N2429
OC41	AC122-125-138-151-173, BC126-260A, NKT211, SK3004, 2N1190-1303-1352-1373-2613-5022
OC41N	AC122-125-151-173-191, BC126-260A, NKT281, 2N1191-1352-1371-1384-1991
OC42	AC122-125-151-173-191, BC126-260A, NKT281, SK3006, 2N1191-1303-1352-1371-1384
OC42N	AC122-125-151-173-191, BC126-260A, NKT211, 2N1192-1352-1373-1384-1991
OC43/N	AC122-125-151-173-191, BC126-260A, NKT211, SK3006, 2N1192-1309-1352-1384-1991
OC44	AC191, AF125-132-134-137-188-196, AFY15, ASY27, BC126, BSW19-72, NKT211, SFT358, SK3005, 2SB44-101, 2N36-1191/2-1352-1373-1384
OC44N	AF125-135-172, BFX48, BSW19-72, FT1746, NKT674F, SFT358, 2N2411-2635-3324
OC45	AF132/3-137-172-185-187-196, AFY15, ASY26-27, BFX48, FT1746, NKT674F, SK3005, 2SB45-101, 2N36-2411-2635-3324
OC45N	ASY26, BC260A, BSX36-NKT135, 2N1192-1352-1371-1384-2927
OC46	ASY27-46, BC260A, BSX36, AF133, NKT135, SFT226, SK3005 , 2SÃ212, 2N1352-1371-1384-2927

OC47	AF188, AC191, ASY47, BC260A, BSX36, NKT135, SFT227, SK3005, 2SA217, 2N1352-1373-1384-2927
OC57	AC122Y-125-129Y/R, 2SB47, 2N1190
OC58	AC122V-129V, OC624G, SK3003
OC59	AC122 Black-129V, AF129B
OC60	AC122Y-125-129B-151V, AF129B, SK3003
OC65	OC57, ST301, SK3004
OC66	OC58, ST302, SK3004, 2N2429
OC70	AC121-122R-126-151IV, ASY27, BC213-260A, BSX36, OC304I, SK3003, 2N1193-1352-1375-1384-2429
OC71	AC122R-126-151IV/V-191, BC126-213-260A-304, NKT211, OC3041, SFT352, SK30 04, 2N1190-1352-1371-1384-1991-2429
OC71N	AC126, 2N2429
OC72	AC122/3-125/6-131 -151IV-152-162-191, BC126, 213-26 0A, NKT211 , SFT353, SK3003, 2N282-1190-1352-1371-1384-1991-2431
OC73	GFT34, OC75, SK3003
OC74	AC125-151V-152-180-192, BC126-213-260A, NKT264, SK3003, 2CY20, 2N1193-1352-1375--1501-1924-1991-2431-2613
OC74N	AC117-124-128-152-180-192, BC126-260A, NKT229, 2N1192-1352-1373-1991-2613
OC75	AC122G-15/IV/V/VI-116-173-192, BC126-213-260A, NKT219, OC304III, SK3003, 2N466-2429-2613-2801-2927-3638
OC75N	AC122-125-151-173-191, BC126-260A, NKT211, SK3004, 2N1191-1352- 1371-1384-1991
OC76	AC117-125-132-151IV-180-184-192, ASY48IV, BC126-213-260A, NKT211 SK3004, 2N394-1191-1352-1371-1384-1991
OC77	AC184-192, ASY81, ACY17, NKT219, SK3004,, 2N24A-284A-2000-2613-2801-2927-3638-4030
OC77N	AC184, ACY24, ASY48-77-81, BCW86, NKT302, 2N398-1998-2000-2837-4030-5151
OC79	AC123-125/6-151IV/V-152-184, ACY24, BC260A, BCW86, NKT302, SK3004, 2N398-1998-2387-2431-2837-5151
OC80	AC121VI-124-126-131-152-162-180-184-192, BC260A, BSX36, 2CY21, OC318, NKT304, 2N1189-1998-2000-2837-2927
OC80A	AC126
OC81	AC117A-153-184, OC318, SK3004
OC83	AC117-124K-131-162-180-192, 2CY21, NKT302, BC260A, BSX36, 2N1189-1924-1999-2000-2837-2927
OC110-120	AC126, 2N2429
OC122	AC117-124/K-128-131- 52V-153/V-180-192, ASY80, BC231-260A, BSX36, 2SB222, 2N467-1998-2837 4106-5023
OC123	AC184-192, ACY24, ASY24-48V-77-81, NKT302, 2N241-1998-2001-2801-2837-4030-5040

OC130	AC126, SK3005,
OC139	AC127-130, AF180-188-192, ACY24, ASY28, BSW82, SFT259, NKT302, SK3011, 2SC89, 2N398-2837-5151
OC140	AC127, ASY29-75, BSW82, BSX12, NKT734, SFT260, SK3005, 2N1306-1605A-1808-1994-2430-5135
OC141	AC127, ASY29-75, BSW82, BSX12, NKT734, SK3011, SFT261, 2N1308-1605A-1808-1994-2430-5135
OC169	AC127, AF132-133-135-185-197, BSW72, BSX12, NKT734, SK3006, 2N346-1308-1605A-1808-2430-5135
OC170	AF132-133-137 185-196, BSW72, BSX48, NKT674F, SFT357P, SK3006, 2G4015, FT1746, 2N370-2273-2653-2873
OC171	AF136-172-178, BSW72, BFX48, FT1746, NKT674F, SFT357P, SK3006, 2G4015, 2N2273-2495-2635-2873
OC171M	AF124/5-131-134/5-195, 2SC135, 2N299
OC171V	AF114-124-130-134/5-194, BF342, SK3006, 2SA235, 2SC135, 2N396-384-4035
OC200	AC122/30R, BC153-177-204/5-178-212-261/2-308VI, BSX29, NKT20329, MPS6516/7, FT5040, SK3114, 40319, 25A565, 2N328-3638-4058-4126
OC201	AC122, BC177V-178-205-213-261/2-291D-308VI, FT5041,, MPS6517, NKT20339 2SA565, 2N3638-4059-4126
OC202	ASY26/7, BC179-206-263-291D, FT5041, NKT20329, SK3114, 2SA155, 2N799-3649-4059-412.5.
OC203	BC177-204-261-291D, NKT20339 SK3114 2N3644-4060-4126-4917
OC204	BC179-206-291D, SK3114 2N3250-3671-3962-4037-4142
OC205	BC177-204, BFX48, NKT10339 SK3114, 2N3644-3962-4060-4126-4917
OC206	BCY33, BFX48, ME0401, SK3114, 40406, 2N3250A 3702-3962-5366
OC207	BC143-212, BCY32, ME0402, MPS3702 SK3114, 40406, 2N3250-3702-3962-5367
OC300	AC128-131/2-152/3-184
OC302	AC122-125/6-151-153-172/3, BC260A, BFX48 FT1746, NKT613F, 2N2273-2429-2635-2873
OC303	AC122R-151IV-173-191, BC126-260A, NKT211, 2SB75, 2N1191-1352-1371-1384-1991
OC304	AC122-151-191, BC126-260A, NKT229, OC71-604, SK3004, 2N1193-1375-1991-2613
OC304I	AC122/R-125-132-151IV-163, SK3004, 2S39, 2SB77, 2N220
OC304II	AC 122Y-123-125-151V-163, 2SB77A-101, 2N238
OC304III	AC122G-125-128-151V/VI-163, 2SB77B-219, 2N266
OC305	AC122-151-173-191, BC126-260A, NKT229, GPT22/15, SFT353A/B, SK3004, 2N1192-1373-1991-2613
OC305N	AC125

OC305I	AC122/V-126-128-151VI/VII, SK3004, 2SB77
OC305II	AC122W-125R-126-151VII, SK3004, 2SB77A
OC306	AC122-123-125R-151R-173-191; BC126-260A, NKT219, 2N1191-1352-1991-2613
OC306I	AC125R-150-151R-161/2, 2SB73-220, l2N191
OC306II	AC125R-150/Y-151R-161-162, 2SB73A, 2N191 -220
OC306III	AC125R-128-150/G-151VI,R-161, 2SB73B-220, 2N191
OC307	AC122-152IV-173-184-191, ASY76, BC126--260A, NKT211, SK3004, 2N1190-1352-1371-1384-1991
OC307I	AC132-151IV, ASY76, SFT321
OC307II	AC132-151IV, ASY76, SFT322
OC307III	AC132-151V, ASY80, SFT323
OC308	AC122-125-132-151/V-184, ASY76, BC126-260A, NKT211, SK3004, 2SB156A, 2N1191-1352-1371-1924-1991-2613
OC309	ASY48IV-81, ACZ10, AC181, BCW86 NKT302, 2N398-2837-4030-5131
OC309I	AC128, ACY24K, ASY48IV-77, 2N1924
OC309II	AC128, ACY24K, ASY48IV-77, 2N1925
OC309III	AC128, ACY24K, ASY48V-80, 2N1926
OC318	AC121IV-152-180-192, BC260A, BSX36, NKT219, SK3003, 2N1924-1998-2000-2431-2836-2927
OC320	OC60
OC330	OC65-622, SK3004
OC331	AC129R, SK3006
OC340	OC624, SK3004
OC341	AC129Y, SK3004
OC342	OC624G, SK3004
OC343	AC129Blue, SK3004
OC350	AC126, SK3004
OC351/61	SK3004
OC366	OC65-623, SK3004
OC362/3/4	SK3004
OC390	AF126-136-185-196, AFY15, BC260A, NKT211, SK3008, 2N1193-1373-2613-5040
OC400	AF136-185-196, BC126-260A, NKT211, 2N1192-1373-2613-2927
OC410	AF136-185-196, BSY27, BC126-260A, NKT211, SK3005, 2N1193-1375, 2613-2927
OC430	BC178IV--179-206-231B-263-291D -309, BCY33-34, BSY40 SK3114, 2N3250-3671-4037-4142
OC430K	BCZ11, SK3114
OC440	BC177-178VI-204/5-213-261/2-291D-307, MPS6518, SK3114, 2SA515, 2N3250A-3702-4037-4142
OC443/5	BC117-178VI-204-205-213-261/2-291D-307, MPS6518, SK3114, 2SA565, 2N3250A-3671-3702-4037
OC443K	BCY27-29, BCZ10, SK3114
OC445K	BCY19, BCZ12, SK3114
OC449	BC117/VI-204/5-213-261/2-291D-307, BCY67, BCW56A, MPS6518-6563, SE8540, SK3114, 40406, 2SA565, 2N1132
OC449K	BCY29, BCZ12, SK3114
OC450	BC205-213-262-291D, BCY70, BSV68, ME0461, MPS6518-6523, SK3114, 2SA565, 2N3638A-3962

OC450K	BCZ12
OC460	BC178VI-179-205/6-213-262/3-309. BFX48, MPS6518, NKT20339, SK3114, 2SA565, 2N3644-4060-4126-4917
OC463	BC178VI-205-213-262, MPS6518, 2SA565, SK3114
OC463K	BCY28, BCZ11, SK3114
OC464	BC179-206-263-291D-309, BCY39, 40406, 2N3250A-3671-3702
OC465	BC178VI-179-205/6-213-262/3-291D-309, MPS6518, SK3114, 40406, 2SA565, 2N3250A-3671-3703
OC465K/66K	BCY28, BCZ11, SK3114
OC466	BC178VI-205-213-262, MPS6518, SK3114, 2SA565
OC467/8/9	BC178VI-204/5-213-261/2-291D-307, BCY38, NKT20319, SK3114, MPS6518, 2SA565, 2N3644-4059-4125/6-4916
OC467K	BCY28, BCZ11, SK3114
OC468K/69K	BCY18, BCZ11, SK3114
OC470	BC117-178VI-205-261/2-291D-307, MPS6518-6523, SK3114, 40406, BCY38, 2SA565, 2N3703
OC470K	BCY17, BCZ10-11, SK3114
OC480K	BCY20
OC601	OC70, 2N2429
OC602	AC122R-151IV-152, OC304I, SFT351, SK3004, 2SB459, 2N238-2429
OC602Spez	AC125-131-151VI-152/IV-180, OC318, 2N1924
OC603	AC125-130-151R-151IV-161, SK3004, 2SB451, 2N207B-238-2429
OC604	AC151, SK3004, 2SB451, 2N238
OC604S	AC131
OC604Spez	AC117-128/K-131-152/3-180K, 2S32, 2N610
OC612	AF133-184-197, AFY15, SK3008, 2SA240, 2N642-4034
OC613	AF132-185-196, AFY15, SK3008, SFT316, SK3008, 2SA240, 2N113-641
OC614	AF185-195, SK3008, 2SA156, 2N110
OC615	AF178, SK3006, 2N2495
OC615M	AF125-131/2-134/5-195, OC615, 2SA76, 2N346-4035
OC615V	AF124/5-130-132-134/5-178-194, 2SA76, 2N1110-2495
OC622/3	OC65
OC624	OC66
OC810/1	2N2429
OC6015	2N2495
OD16	AD130-138-149-153, NKT453, SFT265, 40022, 2N178-250A-1529
OD20	AD163, ASZ15, NKT401, SFT268, 40050, 2N1041-1073B-1100
OD22	AD149-152, ASZ16, MP2061, NKT420, SFT266, 2N1545-2869
OD23	AD133-149-152/3, MP2062, NKT404, SFT265, 40051, 2N250A-1540
OD24/5	AD149-150-153, MP2061, NKT404, SFT265, 40421, 2N251A-1545

OD26/7	ASZ16, AUY21-32, MP2063, NKT452, SFT265, 40022, 2N251A-1529
OD28	ASZ15, AUY28-30-34, MP2060, NKT401, SFT267, 40021, 2N1041-1531
OD29/35	ASZ16, AUY21-31, MP2061, NKT404, SFT266, 40421
OD30	AD148-162, MP2061, NKT453, 40022, 2N251A-1539
OD36	ASZ18, AUY22-30, MP2061, NKT403, SFT268, 40421, 2N1540-1550
OD603	AD131IV/V-138/50, ASZ18, AUY22-30, MP2061, NKT403 SFT268, TI3029 40421, 2N1540-1550-2066- 2836
OD604	2N2836
OD605	OC26, 2N2836
OD650	AD138, ADZ11
OD651A	AD133-138, ADZ11
OX3003	AC126- 2N2429
OX3004	AC128, 2N2431
OX4001	AF185
P1027	E270
P1028/69E	E271
P1086E	E174
P1087E	E175
PADT23	AF185, SK3005
PADT24/5/8/30	AF178, SK3006, 2N2495
PADT31	AF178, SK3006, 2N2495
PBC107	BC237
PBC108	BC238
PBC109	BC239
PET1001	BC207, BCY59A-66, BSX45VI-51A, ME1001, MPS6575, NKT10339, SE1001, 2N929
PET1002	BC207, BCY59B-66, BSX45X-52A, ME1002, MPS6575 NKT10439, SE1002, 40245, 2N930
PET1075	BF156-177-257-297, ME1075, SK3039, 40360, 2N1990N-3114
PET1075A	BF118-156-178-257/8, ME1100, MM3001, SE7002, 40349, 2N3114
PET2001	BFX62, BFY39I, BSX51A-69-81, ME2001, MPS6575 NKT10339, SE2001, 40244, 2N929
PET2002	BFX59, BFY39II, BSX52A-69-81, ME2002, MPS6575, NKT10439, SE2002, 40245, 2N930
PET3001/2	BFX59, BFY39-66, BSX92, BSY95A, ME3001, MM1501, NKT16229, SE3001, TIS62, 40480, 2N2368-3011
PET3702/1	BC328-360VI, BSV15VI, BSX36-GI3702, ME0401, MPS3702, NKT20329, 40406, 2N2906-3702-5447/8
PET3704/5/6	BC337-340VI, BSX30-45VI, BSY51, ME6001/2, MPS3704/5, GI3704/5, NKT10339, SE6001/2, 40317, 2N5449
PET3903	BSX33-46VI, BSW84, BC184, ME6101, 40407, 2N2221-3903
PET3904	BC184, BSX33-46X, BSW85, ME6102, 40407, 2N2222-3904

PET3905/6	BSX36. BSW74. BSV16V1 BC214. ME0401. 40406. 2N906. 3905
PET4001/2	BC168A-183-208. BCY58A. ME4001 MPS6574. NKT10419. SE4001. SK3024. 40398. 2N3706
PET4003	BC168C-183-208. BCY58C. ME4003. MPS6574. NKT10519. SE4003. 40397. 2N3706.
PET4058/9/60	BC153-182-205. BCY78VI. GI3703. MPS3703. NKT20229. SE8540. 40406.
PET4061/2	BC153-182-205. BCY78A. GL3702. MPS3702. NKT20329. SE8540.
PET4123	BC207. BCY59. ME9001. BSX88A. NKT20339. SE8040. 2N4123-4140.
PET4124/5/6	BC208. BCY58. BSX88A. ME9002. NKT20229. SE8040. 40407. 2N4124-4141. •
PET6001/2	BC222-337-340. ME6001/2. MP5. 3704/5. GI3704/5. NKT10339. SE6001/2. 40451/2. 2N3416/7-5449/50
PET8000/1/2/3/4	BC183-289. BCY56-59-66. BSX51/2-54. MPS6575. 40456. 2N3566.
PET8005/6/7	BC184-208. BCY58. MPS6574. NKT10339. 40311. 2N3565.
PET8101	BSV90. BSY19-63. EN744. ME8101. NKT16229. 2N744-3013-3839-4264.
PET8200	BC168-172-183-208. ME4001. MPS 6574. NKT12329. 40311. 2N3565.
PET8201	BC167-172-183-207. ME8001. MPS6575. NKT10339. SK3039. 40456. 2N3566
PET8202/3	BC168-172-183-208-288. MPS6574. NKT10439. 40311. 2N3565/6.
PET8203	BC169-173-183-209-239. MPS6575. NKT10519. 40456. 2N3565/6.
PET8250/1	BC222-337-340. GL3702. ME6001. MPS3702. NKT10439. SE6001. 40451. 2N3416. 5450.
PET8300/1/2/3/4	BC204. BCY79. BSX36. GI3703. MPS3703. NKT2039. SE8540. 40406. 2N5448.
PET8350/1/2/3	AC117-124-128-153-180-192. BSX36. NKT302. 2N1998-2000-2837-3638-40
PL1091	E304
PL1092/3/4	E305
PT4416	OC16
Q6/7/8	AC128. SK3004. 2N2431
RF1	AF185
RR83/7/117	AC126. 2N2429
RR160/1/2	AF185
RRJ14/20/34	AC126. 2N2429
S1042	BC177-237-257V1-307V1. MPS6516. 2N3703
S1211N	E201
S1212N	E201
S1213N	E201
S1214N	E201
S1215N	E201
S1216N	E202
S1221N	E201
S1222N	E201

S1223N	E201
S1224N	E201
S1225N	E201
S1226N	E202
S1231N	E201
S1232N	E201
S1233N	E201
S1234N	E202
S1235N	E202
S1236N	E202
S1241N	E304
S1242N	E304
S1243N	E304
S1244N	E211
S1245N	E113
S1246N	E112
S2048	BC177-237-257A-261-308VI, MPS6516, 2N3702
S2049	BC140-337-337/16, 2N2222-3704
S2050	BC107-147A-267A-337A, MPS6566, 2N3710
S2292	BC177-237-257A-261-308VI, MPS6516, 2N3702
S15649	MPS6520, SK3122
S15650	MPS6515, SK3117
S15657	MPS3563, SK3117
SB100	AF185, SK3008
SC12	AC126, SK3004, 2N2429
SC107	BC182L
SC108	BC183L
SC109	BC184L
SDF500	E400
SDF501	E400
SDF502	E400
SDF503	E401
SDF504	E413
SDF505	E400
SDF506	E400
SDF507	E400
SDF508	E401
SDF509	E413
SDF510	E400
SDF511	E400
SDF512	E400
SDF513	E401
SDF514	E413
SDF1001	E108
SDF1002	E109
SDF1003	E110
SDT1860	2N4281-4282-4283
SDT1861	2N4051-4279-4280
SDT1862	2N3316-4048
SDT1900	2N4051-4052-4053
SDT1961	2N4279-4280-4281
SDT9901	2N4301
SDT9902	2N4301

SDT9903	2N4301
SDT9904	2N4301
SE1001	BC118, BF173-194-224, ME1001, MPS3693, NKT10339, SK3018, TIS87, 2N3693, 40235
SE1002	BC134-182A-171A-207A-237A, BF194-237, ME1002, MPS3694, NKT10439, SK3018-3020, 2N3694
SE1010	BF22F-165-184-189-194-235-237, BFY19, ME8101, MPS3693-6514, NKT16229, SK3018/9, 2N3478-3564
SE2001	BC129-147A-172-207-223, BF154-185-224-235, BFY19, ME2001, MPS6512, NKT35219, SK3020, TIS98, 2N2708-3691
SE2002	BC129-147-172-207-223, BF154-224-234-254, BFY19, ME2002, MPS6513, NKT35219, SK3117, 2N2708-3692
SE3001	BF185-194-195, BFX59-66-73-89, BFY19-88, ME3011, MPS3563, NKT16229, SK3018, TIS62, 2N918-2857-3015-3563-3866
SE3002	BF185-194-195-357, BFX59-73, BFY19-88-90, ME3011, MPS3563, NKT16229, SK3018, TIS62, 2N918-2857-3563-3600
SE3005	BF271-357, BFX62, BFY90, ME3002, SK3117-3018, 2N2616-2857-3839
SE3819	E203
SE4001	BC107-129-132-173B-182-209B-237A-239B, BF357, GI3708, ME4001, MPS6514-6575, NKT10519, SK3020, TIS97, 2N2708-2921, 2SC458
SE4002	BC107-129-132-173C-182-184A-209C-239C, GI3708, ME4002, MPS6576, NKT10519, SK3020, TIS97, 2N5088-2708
SE4010	BC108B-114-131-173C-183-184C-209C-239C, GI3708, ME4010, MPS6576, NKT10519, SK3020, TIS97, 2N2708
SE4020	BC209C, BCY65E/X-79C, BFR16, SK3117, 2N760A-929A
SE4021	BC209C, BCY56-59C-66, BFR17, MPS-A18, SK3117
SE4022	BC209C, BCY59C-65E, BFR16, MPS-A18, SE4021, SK3117
SE5001	BC127-163-167-173-225-237, ME5001, MM800, MPM5006, NKT35219, SK3019, 40238, 2N3337-3588, 2SC464
SE5002	BF127-164-167-173-225-237, ME5001, MPM5006, MPS6544, NKT35219, SK3018, 40239, 2N3338-3689
SE5003	BF162-167-173-225-237, ME5001, MPS6544, NKT35219, SK3018, 40237, 2N3689-5184
SE5006	BF162-225, ME5001, MPM5006, MPS6546, NKT35219, 40237, SK3018, 2N3691-5184
SE5020	BF121-173-200-224-237-311, BFX59, MPS6568, NKT10519-35219, SE3021, SK3018, TIS84, 2N2708-3690
SE5021	BF173-237, MPS6568, SK3018, TIS108

SE5022	BF121-173-200-224-237-311, BFX59, MPS6568, NKT10519, SK3117, TIS108, 2N2708-3690
SE5023	BF127-167-225-310, MPS6569, NKT35219, SK3039, TIS108, 2N3338-3689-40239
SE5024	BF127-167-225-310, MPS6570, NKT35219, SK3018, TIS108, 2N3338-3689
SE5025	BF123-173-224, BFY79, MPS6544, SK3018-3024
SE5050	BF224-225-235-240-310, BFX18-60, SK3018
SE5051	BF224-235-240-255-310, BFX18-60, SK3117
SE5052	BF121-200-225-311-314, BFW64, BFX59, MPS6568, SK3039
SE5055	BF200-225-314, BFW63, BFX59, ME5001, MPS6569, SK3122
SE5056	MPS-H32
SE6001	BC107-107A-182A-237A, MPS6512-6575, SK3020, 2N2921-3242A-3566-2SC458
SE6002	BC107-107B-129-182B-183, MPS6514-6576, SK3020, 3242B, 2N3566
SE6020	MPS/A05, SK3024
SE6020A	BC141/10-341/10, BFW71, BFY44, MPS-A05, SK3024, 2N460A-760A-4226
SE6021	MPS/A06, SK3024
SE6021A	BSW65, BSY56, MPS-A06, SK3024, 2N4963
SE6022	MPS/A05, SK3024
SE6023	MPS/A06, SK3024
SE7001	BF110-111-114-117-119-140-157-178-186-257-258, MPS/U04, SK3010, TIS100, 2N3114-3712-4068
SE7002	BF110-111-117-140-156-178-179A-186-258, MM3008, MPS/U03, SK3045, TIS100, 2N3712-5184
SE7010	MPS/U04, SK3045
SE7015	BF117-178, BFX68, MPS/L01, SK3021, 2N3498
SE7016	BF111-120-174-186-257-258, 2N3114-3500-4068-5550, SK3045
SE7017	BF111-120-186-258, MM3009, SK3045, 2N5551
SE7055	BF120-258, MPS/U10, SE7055, SK3045, 2N4927
SE7056	BF259, MPS/U10, SK3045, 2N3742
SE8001	BC140/6-211, BCY51, BSY34, MPS/U02, SK3024, 2N2217-2218-2218A, 2SC479
SE8002	MPS/U02, SK3024
SE8040	MPS6560, SK3122
SE8041	MPS/U02, SK3024
SE8042	MPS/U02, SK3024
SE8540	MPS6562, SK3025
SE8541	MPS/U52, SK3025
SE8542	MPS/U52, SK3025
SES3705	BC337
SF115	BF595
SF167	BF596
SF173	BF597
SFT101	AC125-126, 2N2429
SFT102	AC125-126, 2N2429
SFT103	AC126, 2N2429
SFT105	AC126, 2N2429

SFT106	AF116-126-185
SFT107	AF116-128-185
SFT108	AF115-125-185, SK3087
SFT109	AF125-126, 2N2429
SFT111	AC125
SFT112	AC132
SFT113	AD149, OC26, 2N2830
SFT114	ASZ15-17
SFT115	AF116-126-185
SFT116	AF115-125-185
SFT117	AF114-124-178, 2N2495
SFT118	AF114-124-178, 2N2495
SFT119	AF116-126-185
SFT120	AF115-125-185
SFT121	AC128-132, 2N2431
SFT122	AC128-132, 2N2431
SFT123	AC128-132, 2N2431
SFT124	AC117-128-131-152-153-153VI-180-192, BSX36, NKT304, 2N249-1998-2000-2106-2431-2801-3638-4030-5023, 2SB226-370A
SFT125	AC117-126-128-131-153-153V-153VI-174, 2N223-249-2431-4106, 2SB222-370A
SFT126	ASY26, 2G138 139
SFT127	ASY26, 2G138 139
SFT128	ASY27; 2 G140-141
SFT130	AC117-128-128K-151V-151VI-153K, ST172, 2N386-2431
SFT131	AC117, AC128-128K-131-153K-153V-153VI, 2N223-2431,2SB227-415
SFT142	ASY80
SFT143	AC124-128-153
SFT144	AC124-127-128-153, 2N226, 2SB222
SFT145	AC124-127-128K-153K
SFT146	AC124-127-128-128K-153K
SFT150	ASZ15-16
SFT151	AC125-126, 2N2429
SFT152	AC125-126, 2N2429
SFT153	AC126
SFT162	AF118, 2N2207, 2SA76
SFT163	AF121-182-200, AFZ12, SK3006
SFT170	AF109-109R-139-180, BFW20, BSW72, NKT613F, 2N2273-3127-3324-3588-4916
SFT171	AF102-106-190, BFX48, BSW72, FT1746, NKT613F, 2N2273-3323, 2SA230
SFT172	AF102-106-190, BFX48, BSW72, NKT603F, 2N2411-2635-3324, 2SA230
SFT173	AF121-200-201, BFX48, BSW19-72, NKT677F, 2N705-2412-3127-3324-3883; 2SA239
SFT174	AF102-121-121G-202, BFX48, BSW19-72, NKT677F, 2N705-2412-3127-3324-3883, 2SA239
SFT184	AC127-179, SK3010, 2N1304-2430, 2SD96
SFT186	BF110-114-117-140-178, BFY43, SK3045
SFT186N	BF110-117-178-258, 2SC856

SFT187	BF110-114-117-178, BFY43, SK3045
SFT187N	BF117-178, 2SC856
SFT190	AD132-138/50-149, ADY23, ADZ11-12, AUY21-33, NKT453, SFT190, SK3009, TI3030, 2N250A-514-2065-2869-2870-3616
SFT191	AD131-138/50-149, ADZ11, ASZ18, SK3009, TI3029, 2N2870-3616, 2SB471
SFT192	AD138/50, ADY23, ADZ11, AUY21-33, SFT192, SK3009, 2N251A-513-2869
SFT206	ASY14-26, BC126-260A, NKT211, 2N1191-1352-1371-1384-1991-3486A
SFT207	ASY12-14-26, BC126-260A, NKT211, 2N1191-1352-1371-1384-1991-3486A
SFT208	ASY13-14-27, BC126-260A, NKT219, 2N1192-1371-1991-2447-2613-3486A, 2S13
SFT211	ASZ18
SFT212	AD130-130III-138-149-153, NKT451, SK3009, TI3028, 2N83A-456A-1073-1539-2869-3615, 2SB242
SFT213	AD131III-138/50-148-149-153, ASZ16, AUY28, NKT403, SK3014, TI3029, 2N176-257-514A-1540-2810-2869-3616, 2SB242
SFT214	AD131-131III-131IV-138/50, ADZ11, ASY17, AUY28, NKT403, SFT234, SK3014, TI3029, 2N513A-1540-2869-2870-3616
SFT221	AC122/30-124-125-128-131-153-173-191, ACY23-23V, ASY14-26, BC260A, BSX36, NKT219, SK3004, 2N1303-1307-1189-2000-2613-2836-2927
SFT222	AC122/30-124-125-128-131-153-173-192, ACY23-23U, ASY14-26, BC257-260A, BSX36, NKT219, SK3004, 2N1189-2000-2613-2836-2927, 2SB219
SFT223	AC117-124-128-131-153, ASY14-26, BC126-260A, HJ15, NKT211, SK3005, 2N215-1191-1352-1373-1384-3486
SFT226	AF101, ASY14-26, BC126-260A, GFT45, NKT211, OC612, SK3005, 2N269-1191-1305-1352-1371-1384-1991, 2S40.
SFT227	ASY14-26-27-30, ASZ10, BC126-260B, NKT211, SK3005, 2N978-1192-1307-1373-2613
SFT228	ASY14-27-30, ASZ10, BC126-260A, NKT211, SK3005, 2N978-1193-1309-1375-2613-3486A
SFT229	ASY13-27, BC126-260B, NKT211, SK3005, 2N1193-1309-1375-1991-2613
SFT232	AC122-124-128-151-153-173-191, BC126-260A, CTP1104, NKT211, OC104, 2N176-1191-1352-1371-1384-1991, 2SB41
SFT233	AC128, ACY24, ASY48-77-81, BCW86, NKT217, 2N390-1021-1030-5131
SFT234	ACY24-48, ASY48-77-81, BCW86, NKT217, 2N398-1924-4030-5131
SFT237	AC160-160B, ACY32-32V, BC260A, BSX36, NKT302, 2N1189-1998-2001-2836-2927

SFT238	AD138-138/50, ADY23, ADZ11, ASZ16-17, AUY21-21III-28-31-33, NKT405, SK3009, TI3029, 2N101-513A-2869-3614-3615, 2SB242
SFT239	AD138/50, ADY25, ADZ12, ASZ16, AUY21-21III-28-31-32-34, NKT403, SFT239, SK3009, TI3028-3029-3030, 2N513B-2870-3615
SFT239Y	ASZ16-18, CDT1311
SFT239GO	AD149, ASZ17, CDT1311, 2N352, 2SB86
SFT240	ADY25, ADZ12, ASZ18, AUY22-28-30-32-34, NKT401, SFT240, SK3009, TI3029-3030, 2N513A-2870-3613-3615
SFT240GO	AD131, ASZ15-18, 2N157A, 2SB87
SFT2400	AD105, ASZ18, GFT3408, 2N268
SFT241	AC128-131-131/50-152-152IV, ASY48, BSX36, NKT302, 2N44-1191-1924-1991-1997-1998-4030, 2SB224
SFT242	AC117-128-131-131/30-152-152IV-161-162, ASY48, BC261, BSX36, NKT302, 2N284-1925-1998-2000-2801-3638-4030-5040, 2SB89
SFT243	AC128-192, ACY24, ACZ10, ASY14-48-48IV-81, BCW86, NKT302, 2N24A-34-86A-241-1998-2000-2801-3638-4030-5131-2SB89
SFT250	AD132-132III, ADY23, ADZ11, ASY41, ASZ18, AUY28-29-30-31, NKT406, SFT250, SK3009, TI3029 2N268-301-514B-2870-3615, 2SB340
SFT251	AC125, ASY26, SK3005
SFT252	ASY26, SK3005
SFT253	AC122G-125-163, 2N4030, SK3005
SFT260	ASY74, OC140, SK3011, 2N1090, 2SC90
SFT261	ASY75, SK3011 2N1091, 2SC91
SFT264	AD132IV-133-138, ADY24, ADZ11, AUY21-28-30, NKT404, SFT264, SK3012, TI3030, 2N514A-2869-3616
SFT265	AD103-133, ADY24, ADZ11, AUY21-28-30, NKT404, SK3012, TI3029, 2N514A-1146-2869-3615
SFT266	AD104, ADY24, ADZ11-12, AUY21-28-30, NKT403, SFT267, SK3012, TI3029, 2N365-514A-1146A-2870
SFT267	AD105, ADY25, ADZ11-12, AUY28-30-34, NKT401, SK3012, TI3029, 2N513A-1146B-2869-3614
SFT268	ADY25, ADZ12, AUY28-30-34, NKT401, TI3030, 2N514B-1100-2870-3613
SFT288	ASY27, BC126, BSW72, NKT219, TI3030, 2N1191-1309-1352-1371-1491-2613
SFT298	ASY27, BC126, BSW72, NKT219, SK3011, 2N1191-1308-1352-1371-1991-2613
SFT306	AC117-126-128-163-184-192, AF101-121-185, AFY15, NKT304, SFT307, SK3004, 2N218-506-1192-1373-1997-2836-2905-2927-4030, 2SA12, 2SB459
SFT307	AC117-121IV-128-163-184-192, AF121-185-187, AFY15, BSX36, GET45, NKT304, SK3005, 2N409-1192-1373-1997-2836-2905-2927, 2SA12

SFT308	**AC117-128-151-163-192, AF101-121-185-190, AFY15, BSX36, NKT302, OC613, SK3008, 2N112-1192-1373-1998-2836-2905-2927-4030, · 2SA15**
SFT315	**AF118, ASZ20, SK3008, 2N2207**
SFT316	**AF127-132-137-138-170-185, BFX48, BSW19-72, FT1746, NKT674F, SK3008, 2N247-2273-2635-2873, 2SA215**
SFT316B	**AF126-132-137-138-197**
SFT316V	**AF126-132-138-198**
SFT317	**AF125-131-135-136-137-170-185-193, BFX48, BSW19-72, FT1746, NKT674F, SK3008, 2N372-1178-2273-2635-2873, 2SA81**
SFT319	**AF126-133-136-170-185, BFX48, BSW19-72, FT1746, NKT674F, SK3005, 2N373-2273-2635-2873, 2SA82**
SFT319G	**AF126-132-137**
SFT319B	**AF126-132-137-138**
SFT320	**AF121-126-131-132-136-137-172-185-190-200, BFX48, BSW72, NKT677F, SK3008, 2N374-705-3127-3323-3371-3883, 2SA83**
SFT321	**AC117-125-128-131-132-152-152III, BSX36, NKT302, OC318, 2N1305-1998-2000-2801-3638-4030-5040; 2SB76**
SFT322	**AC117-125-128-131-131/30-132-152-152IV-192, BC260B, BSX36, NKT302, OC318, SK3004, 2N1189-1998-2000-3638-5040, 2SB76**
SFT323	**AC125-131-132-152-152V-152IV-153-192, BC260B, BSX36, NKT304, OC318, SK3004, 2N408-1189-1998-2000-3638-5040, 2SB78**
SFT325	**AC128-131**
SFT335	**AC125**
SFT337	**AC125R-150-151R-160-161-191, BC126-260A, NKT211, OC306, SK3004, 2N1190-1352-1371-1384-1991-2429**
SFT351	**AC122-122R-125-126-131-151-151IV-191. BC126-260A; NKT211, OC304, 2N63-1191-1352-1373-1991-2429, 2SB78**
SFT352	**AC117-122R-125-128-131-151V-162-192, BSX36, NKT304, OC304, SK3004, 2N591-1189-2000-2429-2613-3638-4030-5040, 2SB120**
SFT352FB	**AC107-150, OC603**
SFT353	**AC122-122gr-125-126-151-151VII-163-191, BC126-260A, NKT219, OC304, SK3004, 2N978-1192-1371-2429-2613**
SFT353FB	**AC107-150; OC603**
SFT354	**AF125-131-132-136-170-185-190-196. BFX48, BSW72, FTI746, NKT674F, SK3008, 2N1110-2273-2635-2873, 2SA156**
SFT357	**AF125-131-133-135-170-178-190-195, BFX48, BSW72, FT1746, SK3008, 2N299-1178-2273-2495-2635-2873, 2SA105-156**
SFT357P	**AF118-125-135, SK3006, 2N2207**

SFT358	AF121-124-125-130-134-135-172-178-194-200, BFX48, BSW74, NKT677F, SK3006, 2N299-1110-2495-3127-3324-3371-3883-4916, 2SA105-156
SFT367	AC117-128-131-152-153-161-192, BSX36, NKT219, 2N1191-1352-1371-1384-1924-4030-5040, 2SB370B
SFT377	·AC127-176-185-186, BSW82, BSX12, NKT734, 2N1308-1605A-1808-2430-5135
SFT522	AC127
SFT523	AC128-132, 2N2431
SK3003	AC126-128, 2N2429-2431
SK3004	AC126-128, 2N2429-2431
SK3005	AF185
SK3006	AF178, 2N2495
SK3007	AF178, 2N2495
SK3008	AF185
SK3009	AD149, 2N2836
SK3010	AC127, 2N2430
SK3012	AD149, 2N2836
SK3013	AD149
SK3014	AD149, 2N2836
SK3015	AD149
SO88	AC128, SK3004, 2N2431
SP8A	AC126, 2N2429
SP8B	AC126, 2N2429
SP8C	AC126, 2N2429
ST5	2N2836
ST28C	AF185, SK3005
ST36	2N2836
ST37D	AF185, SK3005
ST121	AC126, 2N2429
ST122	SK3004, 2N2431
ST123	SK3004, 2N2429
ST124	AC126, 2N2429
ST125	2N2429
ST162	SK3011
ST163	SK3122
ST171	AF185
ST172	AF185, SK3011
ST301	SK3004, 2N2429
ST302	SK3004, 2N2429
ST303	AC126, SK3004, 2N2429
ST722	ZT22
ST723	ZT22
ST724	ZT23
STC1015	BLY17
STC1016	BLY17
STC1024	BLY17
STC1400	BLY17
SU2074	E112
SU2075	E112
SU2076	E231

SU2077	E231
SU1078	E401
SU2079	E42
SU2080	E231
SU2081	E231
SU2098/A/B	E400
SU2099/A	E400
T34A	OC65
T34B	OC65
T34C	OC65
T34D	AC128-132, OC72, 2N281-2431
T34E	AC128-132, OC72, 2N281-2431
T34F	AC128-132, OC72, 2N281-2431
T65	AC126, 2N2429
T1040	AD149, OC26, 2N2836, SK3009
T1041	AD149, OC26, 2N2836, SK3009
T1159	AC128-132, 2N2431
T1360	AF126-185
T1361	AF126-185
T1369	AD149, 2N2836, SK3009
T1375	AF125-185
T1376	AC128 2N2431
T1377	AC128, 2N2431
T1390	AF126-185, SK3006
T1675	AF125-185
T1690	AF125-185, SK3006
T1691	AF124-178, 2N2495, SK3006
T1692	AF126-185, SK3006
T1693	AFZ12
T1694	AFZ12
T1695	AF178, AFZ12, 2N2495
T1696	AFZ12
T1727	AF126-185
T1737	AF125-185, SK3006
T1814	AF125-185, SK3006
T1832	AC126
T1833	AC126
T2024	AC126, SK3008
T2028	AC126
T2030	AC126
T2096	AF139
T2097	AF139
T2379	AF178 2N2495, SK3006
T2384	AF178-185, 2N2495, SK3006
T2399	AF178 2N2495
T2400	AF185
T2478	2N706A
T2696	AF139
T2697	AF139
T2796	AF139
T2797	AF139
T2896	AF139
TF49	AC191, AF101, ASY26-27, BC126-260A, NKT211, 2N218-1911-1307-1352-1371-1384-1991, 2SA12

TF65	AC125-125R-126-151R-160-173-191. BC126-260A, NKT211, 2N978-1192-1373-1384-2429, SK3004
TF65R-BR	AC122-125-151, OC304
TF65B	AC122G-122Y-125-151V
TF65BR	AC122-122V-122W-151VII
TF65G	AC122-122G-122Y-125-151V, OC71-604, 2N36, 2SB101
TF65GR	AC122GR-125-126-151V-151IV
TF65O	AC122R-125-151IV, 2N238-1190, 2SB459
TF65R	AC122-122R-125-151IV, 2N63-238-1190, 2SB220-459
TF65V	AC122-122GR-126-151VI, 2N266, 2SB219
TF65Y	AC122-122R-125-151IV, 2N190-238-1190, 2SB101-459
TF65/30	AC122-125-126-151, OC71-604, 2N2429
TF65/30G	AC122, GET25, OC71, 2N36, 2SB101
TF65/30Y	AC122, OC304/1-6024, 2N189, 2SB101
TF65/30R	AC122-125, OC602R, 2N63, 2SB220
TF65/30V	AC122. OC304/3-604V, 2N266, 2SB219
TF65/60	ASY23-77, . OC77
TF65/M	AC125, OC71
TF65/30M	OC71
TF65/60M	OC77
TF66	AC117-124-128-131-152-180-184-192, ASY70, BSX36, NKT211, OC318, 2N44-1303-2000-2613-2801-3638-4030, 2SB220, 5040, SK3004
TF66I	AC128-131R-152IV, 2N2431-4106
TF66II	AC128-131Y-131G-152V, 2N2431-4106
TF66III	AC128-131G-152VI, 2N610-2431. 2SB2
TF66-30	AC117-124-128-131/30-152-180-192, BSX36, NKT302, 2N59-238-1924-2000-2431-2801-2998-3638-4030-5040, 2SB156-222
TF66-60	AC128, . ACY24 ASY14-48-77-81. ASZ10, BCW86, 2N24A-241-396-2837-4030-5131. 2SB89
TF68	AF126-185. OC44
TF69/30	AC128, 2N2431
TF70	ASY73, OC139, SK3011
TF71	ASY74, OC140, SK3011
TF72	ASY74, OC140, SK3011
TF75	AC128, OC72, 2N2431, SK3004
TF77	AC126-128, OC74, 2N2431, SK3004
TF77/30	AC128, OC74, OD603, 2N2431
TF77/60	ASZ15, OC28
TF78	AD148-149-150-153-155, ASZ16, NKT402, OC80, OD603, TI3030, 2N297A-2836-2869-3615, 2SB86, SK3020
TF78/30	AC128-152, AD148-149-150-152-153-169, ASZ16-18, NKT406, OD603, TI3030, 2N513B-2870-2904-3615, 2SB86, SK3009
TF78/60	AD131-162, AS215 AUY21, NKT406, OD603/50, TI3030, 2N513B-2870-2904-3616, SK3009
TF80	AD149, OC26-30, 2N2836
TF80-30	AD130-149-150-153-162, NKT452, TI3028-3029, 2N513-1195-2138-2836-2869, 2SB83-337, SK3009

TF80-60	AD131-138/50-140-149-150-153. ASZ15-16, AUY21, NKT404, TI3029, 2N268-513A-2139-2870, 2SB86-471
TF80-80	AD132-138/50-149, ADZ12, ASZ15, AUY22-28-31, TI3029, 2N5138-2065-2141-2870)-2SB472
TF81/30	AD149, 2N2836
TF85	AD149, OC26, 2N2836
TF90	AD133-137-138-149, ADZ11, ASZ16, NKT402, TI3027, 2N514-2138-2836-2869, OC26
TF90/30	AD149-150-153, ADZ11, NKT402, SFT265, TI3028 2N514A-2139-2836-2869
TF90/60	AD149, ADZ12, OC26
TF260	BC107-107A-182A-197, 2N2921-3568, 2SC458
THP44	AC128-132, 2N2431
THP45	AD149, 2N2836
THP46	AD149, 2N2836
THP47	ASZ17
THP50	AD149, 2N2836
THP51	AD149, 2N2836
THP52	AD149, 2N2836
TI156	AD130-149-150-153-159, ASZ16, NKT452, 2N250A-307A, 40022, SK3030
TI158	AD130-149-150-153. ASZ16, NKT404, 2N351A-1540, 40421, SK3031
TI158A	ADZ11, AUY21-28-33, NKT401, SFT266, 2N375-1541, 40421
TI159	AD138, ADZ11, AUY21-31, NKT452, SFT264, 2N176-1539, 40050, SK3031
TI160	AD138-149, ADY23-33, ADZ11, AUY21, MP1530, NKT404, SFT266, 2N1530-2869, SK3031
TI161	ADY25, ADZ12, AUY22-28-32, MP1531, NKT404, SFT267, 2N1531-2869
TI162	ADY25, ADZ12, AUY28-32-34, MP1532, NKT401, SFT268, 2N1542-1906
TI363	AF121S-139, BFX48, BSW19-73, NKT613F, 2N499-2273-3324-4916-5354, SK3005
TI364	AF121S, BFX48, BSW19-73, NKT613F 2N2273-3324-4916-5354
TI365	AF121S, BFX48, BSW19-73, NKT603F, 2N2273-3324-4916-5354
TI388	AF121S, BFX48, BSW19-73, NKT613F, 2N2273-3324-4918-5354, SK3006
TI389	AF121S, BFX48, BSW19-72, NKT613F, 2N2273-3324-4916-5354
TI390	AF239-240, BF272, 2N3127-3323-3324
TI391	AF239-240, BF272, 2N3127-3324
TI395	AF239-240, BF272, 2N3127-3324
TI397	AF125-134-192, BFX48, BSW19-72, NKT613F, 2N2273-2873-4916-5354,
TI398	AF125-134-192, BFX48, BSW19-72, NKT1674F, 2N2273-2873-4916-5354
TI399	AF121-192, BFX48, BSW19-72, NKT613F, 2N354-3280-3304-3324-4916
TI399	

TI400	AF240-239-240, BF272, 2N3127-3324, SK3006
TI401	AF239-240, BF272, 2N3127-3324, SK3006
TI402	AF239-240, BF272, 2N3127-3324
TI403	AF239-240, BF272, 2N3127-3324 SK3006
TI407	BFS10, BFX55, BFY88, SE3006, 2N2218A-3553-3866, 40347, SK3019
TI408	BFS10, BFX55, BFY88, SE3006, 2N2218A-3553-3866, 40347, SK3019
TI409	BFS10, BFX55 BFY88, SE3005, 2N2218A-3553-3866, 40347, SK3019
TI412	2N3704, SK3122
TI413	2N3705
TI414	2N3706, SK3122
TI418	2N3711, SK3019
TI480	BFY46-67C, BSX73, BSY53-54, ME6001, 2N1711-2219A-2789, SK3122
TI481	BFR21, BSX47, BSY55, ME8002, 2N1893-3020-4239, 40360, SK3122
TI482	BFX67C, BSX48-75, BSY53, ME6001, MPS3705, 2N3414-3705-3736-3866-4013, SK3024
TI483	BFY67C. BSX48-75, BSY51, ME6002, MPS3702, 2N3705-3736-3866-4013-4954, SK3024
TI484	BFY67C, BSX48-75, BSY51-63, ME6002, MPS3705, 2N3705-3736-3866-4013-4954, SK3024
TI485	BFY66, BSV52, BSX12, BSY62-95, ME9001, 2N706A-708A-3303-5065-5183, SK3122
TI486	BC313, BFR21, BSX23-46/6, BSY55, ME8003, 2N912-1975-3108- 40360
TI487	BC313, BFR21, BSX23-46/6, BSY55, ME8003, 2N911-1974-3108, 40360
TI492	BF173-198-199-271, BFX60, NKT16229, 2N3337-3693-3933, SK3020
TI493	BF167-180-241-271, BFX60, ME3011, NKT16229, 2N3337-3693-3933, SK3020
TI494	BF235-255-271, BFX60, BSY82, ME3011, NKT16229, 2N3337-3692-3932, SK3020
TI495	BF173-197-198-199-271, BFX60, NKT16229, 2N3337-3692-3829, SK3020
TI496	BFR19, BFY34-67, BSY44-53, ME6001, 2N1613-2221-2787, SK3122
TI539	AD149-150-153, ASZ16, MP1550, NKT404, 2N1540-2869
TI540	AD130-149-150-153, ASZ17, MP1551, NKT403, 2N1541-2870
TI1121	BDY26, BU110, DTS410, 2N1722A-5239
TI1122	BDY26, BU110, DTS410, 2N1722A-5239
TI1123	BDY25, DTS,107, 2N1724-4347
TI1124	BDY25, DTS107, 2N1724-4347
TI1125	BDY24, DTS106, 2N1723-3055
TI1126	BDY24, DTS106, 2N1723-3055
TI1131	BDY26, BU110, DTS410, 2N1722A-5239
TI1132	BDY26, BU110, DTS410, 2N1722A-5239
TI1133	BDY25, DTS107, 2N1724-4347

TI1134	BDY25, DTS107, 2N1724-4347
TI1135	BDY24, DTS106, 2N1723
TI1136	BDY24, DTS106, 2N1723-3055
TI1141	BDY26, BU110, DTS410, 2N1722A-5239
TI1142	BDY26, BU110, DTS410, 2N1722A-5239
TI1143	BDY25, DTS107, 2N1724-4347
TI1144	BDY25, DTS107, 2N1724-4347
TI1145	BDY24, DTS106, 2N1723
TI1146	BDY24, DTS106, 2N1723
TI1151	BDY26, DTS410, 2N1722A-5239
TI1152	BDY26, DTS410, 2N1722A-5239
TI1153	BDY25, DTS107, 2N1724-4347
TI1154	DBY25, DTS107, 2N1724-4347
TI1155	BDY24, DTS106, 2N1723
TI1156	BDY24, DTS106, 2N1723
TI3027	AD138-138/50, ADY23, ADZ11, ASZ16, AUY21-22IV-31, MP1534, NKT404, SFT265, SK3009, 2N1539-2869, 2SB339
TI3028	ADY23, ADZ12, ASZ15, AUY22-22III-28-32, MP1535, NKT402, SFT266, 2N1540-2869, 2SB341, SK3009
TI3029	ADY25, ADZ12, AUY22-28-32, MP1536, NKT403, SFT267, 2N1541-2870, SK3009
TI3030	ADY25, ASZ15, AUY24, MP1537, NKT401, SFT268, 2N1542-1906
TI3031	ADY25, ASZ15, AUY22III-28-34, MP1538, NKT406, SFT268, 2N1543-1906, 2SB341
TIP27	BDY26, BU105-110, MJE340, 2N3902-5240, SK3103
TIP29	BD131-148, BDY38-80A, 2N3054-4913-4921, SK3054
TIP29A	BD137-149, BDY20-80B, 2N3055-4914-4922, SK3004
TIP29B	BD30, BDY20-80C, 2N3055-4915-4923
TIP30	BD132, BDY82A, 2N4901-4918
TIP30A	BD138, BDY82B, 2N4902-4919
TIP30B	BDY82C, 2N4903-4920
TIP31	DB131-241, BDY38-80A, 2N3054-4913-4921, SK3054
TIP31A	BDY20-80B, 2N3055-4914-4922, SK3054
TIP31B	BDY20-80C, 2N3055-4915-4923, SK3054
TIP32	BD132-242, BDY82A, 2N4904-4918
TIP32A	BDY82B, 2N4905-4919
TIP32B	BDY82C, 2N4906-4920
TIP33	BDY23-38, 2N3054-4913
TIP33A	BDY20-23, 2N3055-4914
TIP33B	BDY20-24, 2N3055-4915
TIP34	2N4904
TIP34A	2N4905
TIP34B	2N4906
TIP35	BDY23-38, 2N3054-4913
TIP35A	BDY20-23, 2N3055-4914
TIP35B	BDY20-23, 2N3055-4915
TIP36	2N4904
TIP36A	2N4905
TIP35B	2N4906
TIP41	BDY23-38, 2N3054-4913
TIP41A	BDY20-23, 2N3055-4914
TIP41B	BDY20-24, 2N3055-4915

TIP42	2N4904
TIP42A	2N4905
TIP42B	2N4906
TIS14	E305, SK3112
TIS18	BSX12, BSY19-20-21-62, ME9002, MPS3563, SK3019-3039, 2N706A-3303-3426-5065-5183
TIS22	BCY59A-66, 2N929-3694, SK3122
TIS23	BCY59B-66, 2N930-3693, SK3122
TIS24	BCY59B-65E, ME6101, 2N929A-3694, SK3039
TIS25	E400
TIS26	E400
TIS27	E401
TIS34	E305, SK3116
TIS37	BC308VI, BCY78VI, BFX48, BSW21A, 2N3638-4126-5040-5138, 40406, SK3025
TIS38	BC308VI BCY78A, BFX73, BSW21A, NKT16229, SE3005, SK3114, 2N3794-3839
TIS39	BFW17, BFX59, BFY90, BSX28, ME3011, NKT35219, 2N743-918-3011-5187, 2SC890
TIS41	E111
TIS42	E112
TIS44	BFY66, BSX28, BSY17-38-95, ME9001, MPS706, NKT35219, 2N743-744-3011-5187, SK3122
TIS45	BFX12, BSY19-63, ME9001, MPS708, 2N708-3137-3706-4427, SK3122
TIS46	BFX12, BSY19-63, ME9001, 2N708-3303-4264-5065-5183, SK3122
TIS47	BFX12, BSY19-63, ME9001, MPS2369, 2N708-3137-3706-4427, SK3122
TIS48	BFX12, BSY19-63, ME9001, MPS2369, 2N708-3137-3706-5183, SK3122
TIS49	BFX12. BSY19-63, ME9001, MPS2369, 2N708-3137-3706- -4427, SK3122
TIS50	BFX12, BSW19VI-82, BSX29, ME8201, 2N2894A-3546-5055-5292, SK3122
TIS51	BFY66, BSX91, BSY18-19-95A, ME9001, MPS2369, NKT13329, 2N744-5187, SK3122
TIS52	BFX12, BSY19-63, ME9001, 2N708-3137-3706-4427, SK3122
TIS53	BFX12, BSW82, ME8201, MPS3639 SK3114, 2N2894A-3304-4207-4257-5141
TIS54	BFX12, BSW19VI-82, BSX29, ME8201, MPS3640, 2N2894A-3546-5055-5292, SK3114
TIS55	BFX12, BSY19-63, ME9001, MPS3646, 2N708-3137-3303-4427, SK3122
TIS56	BF173-198-199-271, BFX60, NKT16229, 2N3337-3693-3933, SK3122
TIS57	BF173-198-199-271, BFX60, NKT16229, 2N3337-3693-3933, SK3122
TIS58	E203, SK3116
TIS59	E304 , SK3116
TIS60	BFS10, BFW46, BFX55, BFY88, SE3006, 2N2218A-3553-3866, SK3024

TIS61	BFX48-88, BSV15/6, BSW44A, BSX40, BSY59, ME0402, SK3025, 2N1132-2906-3135-4037
TIS62	BF271, BFY90, BFX60, ME3001, NKT16229, 2N918;3337-3693-4427, SK3039
TIS63	BF271, BFX60, BFY90, ME3001, 2N918-2710-3014-4427-4875, SK3039
TIS64	BF271, BFX60, BFY90, ME3001, 2N918-2710-3014-3691-4427, SK3039
TIS68	E400
TIS69	E402
TIS70	E402
TIS73	E111
TIS74	E112
TIS75	E112
TIS78	E203
TIS79	E203
TIS83	BF271, BFX60, BFY90, ME3011, 2N918-2710-3013-3692-4427, SK3122
TIS84	BF125-235-240-254-273C, MPS6566, NKT16229, SE2002, 2N3693-3932, SK3039
TIS85	BF121-234-241-255-273D, MPS6565, NKT12329, SE2001, 2N3691, 40238, SK3039
TIS86	BF167, MM8000, NKT13329, 2N3689-3694-3932, SK3039
TIS87	BF166-173, 2N3399-3693-3932-4072, SK3039
TIS88	E304, SK3116
TIS90	BC313-337-340-6, BFR18, BSX45/6, MPSA05,. 2N2221-2297-2788 3110-4400, 40348, SK3122
TIS91	BC212-327-360/6, BFY67, BSV15/6, MPSA55, 2N2906-3135-3245-3467-3644, SK3114, 40406
TIS92	BC313-337-340/10, BFR18, BSX45/10, MPS/A05, 2N2221-2788-3110-4400, 40348, SK3122
TIS93	BC212-327-360/10, BFY67, BSV15/10, MPS/A55, SK3114, 2N2906-3136-3244-3467-3644, 40406
TIS94	BFS10, BFW46, BFX55, BFY88, SE3006, SK3122, 2N3553-3866, 2SC890
TIS95	BC141/6, BFR21, BFX84, BSW10, BSX46/6, BSY55, 2N912-1975-3108, 40360, SK3122
TIS96	BC141/6, BFR21, BFX84, BSW10, BSX46/6, BSY55, ME8002, 2N911-1974-3108, 40360, SK3122
TIS97	BFS10, BFW46, BFX55, BFY88, MPS/A18, SE3006, 2N3553-3866-2SC890, SK3039
TIS98	BC341/10, BCY66E, MPS/A08, SK3018-3039, 2N929A-2222-3694
TIS99	BC140/10-212, BFR21, BFX84, BSW10, BSX46/6, ME8002, 2N911-1974-3108, 40360, SK3039
TIXM14	AF239-240, BF272, 2N3127-3324
TIXM15	AF239-240, BF272, 2N3324
TIXM16	AF239S-240, BF272, 2N3127-3244-3324
TIXM17	AF239-240, AFY34, BF272, MM1139, TIXM101, 2N711A-3324-3883
TIXM101	AF239S-240-251, AFY34-40, MM1139, TIXM16, 2N3127-3324

TIXS35	E112
TIXS39	BFV17, BFY39, 2N918
TIXS41	E111
TJ320	AC126, 2N2429
TJ363	AF185
TJ364	AF185
TJ385	AF185
TJ386	AF185
TJ387	AF185
TJ388	AF185
TJ389	AF185
TJ397	AF185
TJ398	AF185
TJ399	AF185
TJN1	AC126, 2N2429
TJN1B	AC126, 2N2429
TJN2F	AC126, 2N2429
TJN2FB	AC126, 2N2429
TJN2G	AC126, 2N2429
TJN2GB	AC126, 2N2429
TJN3	AC126, 2N2429
TJN4	AC126, 2N2429
TJN300/2	AD149, 2N2836
TJN300/2A	AD149, 2N2836
TK20B	V6/4R2
TK21B	V6/2R3
TK23	ACY31
TK23A	GET104, NKT217;227
TK23C	NKT217-227, SK3004
TK24	ASY64
TK24B	V6/2R4
TK25	ASY60
TK25B	V6/8R
TK26B	V6/2R3
TK27B	V6/2R4
TK28	ASY63
TK28B	V6/2R4
TK30C	NKT126, 2G301
TK30D	NKT126
TK31	ASY55
TK31C	NKT125, 2G302
TK31D	NKT125
TK33	ASY61
TK34	ASY62
TK35	ASY50
TK36	ASY57
TK37	ASY58
TK38	ASY59
TK40	AC128, ACY30, SK3004
TK40A	GET102, NKT213-223, SK3004
TK40C	AC156, NKT213-223, SK3004
TK41	AC128, ACY27, SK3004
TK41C	NKT214;224, SK3004

TK42	ACY28, SK3004
TK42C	AC156, NKT213-223, SK3004
TK45	ACY29, NKT216;226
TK46	ASY50
TK46C	NKT215-225
TK47C	NKT215-225
TK48C	NKT217-227
TK49	ASY53
TK49C	NKT713, SK3010
TK400A	NKT404
TK401A	NKT401
TK402A	NKT402
TP107A	BC182A
TP107B	BC182B
TP108A	BC183A
TP108B	BC183B
TP108C	BC183C
TP109B	BC184B
TP109C	BC184C
TP251A	BC212A
TP251B	BC212B
TP252A	BC213A
TP252B1	BC213B
TP253B	BC214B
TP3638A	TIS93
TP4123	TIS92
TP4124	TIS92
TP4125	TIS93
TP4126	TIS93
TP4257	TIS53. SK3118
TP4258	TIS54, SK3118
TP4274	TIS48
TP4275	TIS48
TPS6512	TIS92
TPS6513	TIS92
TPS6514	TIS92
TPS6515	TIS92
TPS6516	TIS93
TPS6517	TIS93
TPS6518	TIS93
TPS6519	TIS93
TPS6520	TIS92
TPS6521	TIS92
TPS6522	TIS93
TPS6523	TIS93
TR34	AC122-125-151-173-191, BC260A, BSX36, NKT219, 2N1189-1998-2000-3638-5040
TR43	AC117-128-152-180-191, BC260A, BSX36, NKT219, 2N1189-1998-2000-3638-5040
TR44	AC117-128-162-180-191, BC260A, BSX36, NKT219, 2N1189-1998-2000-3638-5040
TR45	AC117-128-162-180-190, BC126-260A, NKT217, 2N1191-1352-1371-1384-1991

TR320	AC122-125-151-173-190, BC126-260A, NKT211, SK3004, 2N1191-1352-1371-1384-1991
TR321	AC125-151-160-173-190, BC126-260A, NKT211, SK3004, 2N1190-1352-1371-1384-1991
TR323	AC125R-151R-160-173-190, BC126-160, NKT217 SK3004, 2N1192-1352-1373-2613-2927
TR383	AC122-125-151-173-190, BC126-260A, NKT217, 2N1192-1352-1373-2613-2927
TR482	AC125-151-160-173-190, BC126-260A, NKT219, SK3004, 2N1193-1375-2613-2927
TR508	AC122-151-173-191, BC260A, BSX36, NKT219, SK3004, 2N1189-1998-2000-2927-3638
TR650	AC124-128-153-177-180-191, BSX36, NKT219, SK3004, 2N1189-1998-2000-2927-3638, 4030
TR653	AC117-128-152-180-191, BSX36, NKT219, SK3004, 2N1189-1998-2000-2997-3638, 4030
TR721	AC122-125-151-173-190, BC126-260A, NKT211, SK3004, 2N1191-1352-1371-1384
TR722	AC122-125-151-173-190, BC126-260A, NKT211, OC66, SK3004, 2N1191-1352-1371-1384
TR760	AF185, 2N520, SK3007
TR761	AF185, 2N521, SK3007
TR802	AF127
TRC44	AF126-185, ASY26, BC126, BSW72, NKT217, 2N1192-1373-2613-2927, SK3005
TRC45	AF127-185, ASY26, BC126, BSW72, NKT217, 2N1192-1375-2613-2927, SK3004
TRC65	AC125
TRC66	AC126
TRC70	AC125-126, ASY26, BC126, BSW72, NKT217, 2N1193-1375-2429-2613-2927
TRC71	AC125-126, ASY26, BC126, BSW72, NKT211, 2N1191-1352-1371-1384-2429-2927
TRC72	AC128-132, ASY27, BC126, BSW72, NKT211, 2N1192-1373-2431-2613-2927
TRC76	ASY76, OC76
TRC77	ASY77, OC77
TRC360	OC58-65
TRC601	AC125-126, OC70, 2N2429
TRC602	AC125-126, OC71, 2N2429
TS1	AF185, SK3004
TS2	AF185, SK3016
TS3	AF185, SK3004
TS7	AF185
TS8	AF185, SK3032
TS9	AC128, 2N2431
TS13	AC126, 2N2429, SK3004
TS14	AC126, 2N2429, SK3004
TS161	2xAC132-2xOC72
TS162	AC125-126, OC71, 2N2429
TS163	AC125-126, OC71, 2N2429
TS164	AC125-126, OC71, 2N2429, SK3004
TS165	AC126, OC72, 2N281-2429, SK3004
TS166	AC125-126, 2N2429, SK3004

TS176	AD149, OC70, 2N2429, SK3009
TS306	AC128, 2N2836
TS620	OC58-65-66, 2N2431, SK3008
TS621	OC58-65-70, SK3008
TS739	AC126, 2N2429, SK3004
TS739B	AF185
TS740	AC128, 2N2431, SK3004
TZ81	BC184L, BFX17, BFY67A, BSX49-74, BSY54, ME6002, 2N2194-2224-2270-2868-3053; 40347
TZ82	BC183L, BFX17, BFY67A, BSX49-74, BSY54, ME6002, SK3098, 2N2194-2224-2270-2868-3053, 40347
TZ551	BC213L, BFX48-88, BSV15/6, BSX40, BSY59, ME0401, 2N1132-2838-2906-4037
TZ552	BC214L, BFX48-88, BSV15/10, BSX41, BSY59, ME0402,, 2N1132;2838-2906-4037
TZ553	BC214LC, BFX48, BSV15-16, BSX41, BSY41-59, ME0402, 2N1132-2907-3486, 40406
TZ554	BC213L, BFX48, BSV15/6, BSX40, BSY41-59, ME0401, 2N2906-3485-4890, 40406
TZ581	BC214L-327-360/16, BFX48, BSV15/16, ME0402, 2N2907-3136-3251-3906-4037
TZ582	BC213L-327-360/16, BFX48, BSV15/16, ME0402, 2N2907-3136-3251-3906-4037
TZN3702	2N3702
TZN3703	2N3703
TZN3704	2N3704
TZN3705	2N3705
TZN3706	2N3706
TZN3707	2N3707
TZN3708	2N3708
TZN3709	2N3709
TZN3711	2N3711
U1277	E202
U1278	E201
U1279	E201
U1280	E201
U1281	E114
U1282	E305
U1283	E231
U1284	E202
U1285	E201, SK3112
U1286	E203
U1325	E201
U1714	E202
U1715	E203
U1837E	E305
U1897F	E111
U1898E	E112
U1899E	E112
U1994E	E304
U2047E	E300
UC140	E114

UC155/E/W	E114
UC200	E114
UC201	E114
UC210	E210
UC220	E232
UC241	E232
UC250	E111
UC251	E112
UC400	E270
UC588	E300
UC714	E305
UC734/E	E305
UC755	E203
UC756	E202
UC805	E271
UC807	E176
UC2139	E402
UC2147	E402
UC2148	E402
UC2149	E402
V6/1R	NKT164
V6/2R	NKT143-164, SK3005
V6/2RC	NKT164, SK3005
V6/2RM	NKT164
V6/3R	NKT164
V6/3RM	NKT164
V6/4R	NKT143-163, SK3005
V6/4RC	NKT163, SK3005
V6/4RM	NKT162
V6/6R	NKT163
V6/6RM	NKT162
V6/8R	NKT142-162, SK3005
V6/8RC	NKT162
V6/8RM	NKT162
V6/R2	NKT143-164
V6/R4	GET873, NKT143-163
V6/R8	GET874, NKT142-162
V10/15	AC126, 2N2429
V10/30	AC126, 2N2429
V10/50	AC126, 2N2429
V30/20P	AD149, OC26, 2N2836
V30/30P	AD149, OC27, 2N2836
V60/10P	NKT403, SK3009
V60/20PD	NKT403
V60/20P	NKT403, SK3009
V60/20PD	NKT403
V60/20IP	NKT301
V60/30P	NKT403, SK3009
V60/30NP	GET573
V208	AD149, OC26, 2N2836
V308	AD149, OC26, 2N2836
VB701	V10/50A2
VB704	V10/30A2

VB705	V10/30A
VB706	V10/50A
VB707	V6/4R
VB708	V6/8R
VB709	V10/50A2
VB710	V10/50A
WK5457	E202
WK5458	E203
WK5459	E203
X137	2N677B, SK3012
XA101	AF185, OC45, SK3005
XA102	AF185, OC44, SK3005
XA111	AF185, SK3005
XA112	AF185, SK3005
XA131	AF178, 2N2495, SK3006
XA141	AF178, 2N2495
XA142	AF178, 2N2495
XA143	AF178, 2N2495
XA161	AF178, 2N2495
XB102	AC126, 2N2429, SK3004
XB103	OC66-71, SK3004
XB104	AC126, 2N2429, SK3004
XB112	AC126, 2N2429, SK3004
XB113	AC126, 2N2429, SK3004
XC101	AC128, 2N2431, OC72, SK3004
XC131	AC128, 2N2431
XC171	AC128, 2N2431
XFT2	V10/50A
XS101	OC139, V10/25
XS111	V6/2R
Y203A	XS101
Y203A2	XS104
Y203HB	XS121
Y214N2	XC142
Y214N5	XC155
Y214N6	XC156
Y215J1	XA151
Y215J2	XA152
Y217HA1	XA141
Y217HA2	XA142
Y217HA3	XA143
Y221HB1	XA161
Y221HB2	XA162
Y222HB	XB121
Y363	AC126, 2N2429
Y482	AF185
Y483	AF185
Y485	AF185
Y633	AC128, 2N2431
ZJ13	AC128, 2N2431
ZS4	AF185
ZS5	AF185
ZS8	AF185, SK3030

ZS12	AC128, 2N2431, SK3062
ZS15	AC128, 2N2431, SK3063
ZS30	AF185, SK3030
ZS31	AF185, SK3030
ZS34	AC128, 2N2431, SK3031
ZS35	AF185
ZS36	AF185
ZS38	AC128, 2N2431
ZS41	AF185
ZS43	AF185
ZS45	AF185
ZS52	AF185, SK3030
ZS56	AC128, 2N2431
ZS91	AC128, 2N2431, SK3030
ZS109	AF185
ZS110	AF185, SK3099
ZS112	AF185
ZT80	CV7748, SK3122
ZT82	CV7752, SK3122
ZT87	CV7754, SK3122
ZT180	CV7749, SK3114
ZT182	CV7753, SK3114
ZT187	CV7755, SK3114
ZT204	BSY10, SK3018
ZT402	2S731, SK3018
ZT706	BCY34, 2N706, SK3122
ZT708	2N708-2413, SK3122
ZT709	2N709, SK3039
ZT1420	2N1420
ZT1488	DT4111-4112, 2N1488, SK3027
ZT1490	DT4121, 2N1490, SK3027
ZT1708	BFY44, 2N1708, SK3122
ZT2205	BFY26, 2N2205, 2S131, SK3122
ZT2206	BFY27, 2N2206, SK3122
ZT2270	BFY15-16, DT1521, 2N2270, 28019
ZTX107	BC107-182KS
ZTX107A	BC182KAS
ZTX107B	BC182KBS
ZTX108	BC108-183KS
ZTX108A	BC183KAS
ZTX108B	BC183KBS
ZT X108C	BC183KCS
ZTX109	BC109-184KS
ZTX109B	BC184KBS
ZTX109C	BC184KCS
ZTX114	BC114-184KS
ZTX300	BCW10, ZT1300
ZTX301	BCW12, ZT1301
ZTX302	BCW14, ZT1302
ZTX303	BCW16, ZT1303
ZTX304	BCW18, BFR39
ZTX310	BSV23, TIS44, 2N706
ZTX311	BSV24, BSY95A, TIS45

ZTX312	BSV25, TIS46, 2N706A
ZTX313	BSV26, TIS48, 2N2369
ZTX314	BSV27, TIS49, 2N2369A
ZTX320	BFW97, SX1 2N918
ZTX321	SX18
ZTX325	BF357KS
ZTX326	BF357KS
ZTX327	BF224J
ZTX330	BCW20, 2N3707, 3707KS
ZTX331	BC182KS, BCW22, 2N929
ZTX341	BF297, BSV28
ZTX342	BF297, BSV29
ZTX360	TIS116 .
ZTX500	BCW11, ZT1500
ZTX501	BCW13, ZT1501
ZTX502	BCW15, ZT1502
ZTX503	BCW17, ZT1503
ZTX504	BCW19, ZT1504
ZTX510	BSV33, TIS50
ZTX511	BCW21, 2N4058
ZTX530	BCW21, SK3114, 2N4058KS
ZTX531	BC212KS, BCW23, SK3114, 2N2604

New Additions

New Additions

New Additions

New Additions

New Additions

New Additions

New Additions

New Additions

ALSO AVAILABLE

BP1: 1ST BOOK OF TRANSISTOR EQUIVALENTS AND SUBSTITUTES

AUTHOR: B. B. BABANI   PRICE: 50p
ISBN: 0 85934 000 7   80 Pages
Approx. Size: 180 x 105 mm

The most complete transistor equivalents guide. More than 25,000 transistors with alternatives and equivalents are included. Covers transistors made in Great Britain, USA, Japan, Germany, France, Europe, Hong Kong, and includes types produced by more than 120 different manufacturers.

BP2: HANDBOOK OF RADIO, TV, INDUSTRIAL AND TRANSMITTING TUBE AND VALVE EQUIVALENTS

AUTHOR: B. B. BABANI   PRICE: 60p
ISBN: 0 85934 020 1   96 Pages
Approx. Size: 180 x 105 mm

A modern, easy to use, equivalents handbook for amateur and service engineer. Most new and old valves are shown here with their equivalents. More than 18,000 valves from Great Britain, USA, Europe, Japan and the rest of the world are included and a complete C.V. (Military, Naval and Air Force) list with full commercial equivalents is also provided in a most convenient form.

BP40: DIGITAL IC EQUIVALENTS AND PIN CONNECTIONS

AUTHOR: ADRIAN MICHAELS   PRICE: £2.50
ISBN: 0 85934 044 9   320 Pages
Approx. Size: 180 x 127 mm

Shows Equivalents and Pin Connections of a popular user-orientated selection of Digital Integrated Circuits.

Also shows details of Packaging, Families, Functions, Country of Origin and Manufacturer.

Includes devices manufactured by Fairchild, Ferranti, Harris, ITT, Motorola, National, Philips, R.C.A., Signetics, Sescosem, SGS-ATES, Siemens, SSSI, Stewart Warner, AEG-Telefunken, Texas Instruments, Teledyne.

Compansion volume to book No. BP41: LINEAR IC EQUIVALENTS AND PIN CONNECTIONS.

An invaluable addition to the library of all those interested in Electronics, be they amateur or professional.

BP41: LINEAR IC EQUIVALENTS AND PIN CONNECTIONS
AUTHOR: ADRIAN MICHAELS   **PRICE: £2.75**
ISBN: 0 85934 045 7   320 Pages
Approx. Size: 180 x 127 mm

Shows Equivalents and pin connections of a popular user-orientated selection of Linear Integrated Circuits.

Also shows details of Families, Functions, Country of Origin and Manufacturer.

Includes devices manufactured by Analog Devices, Advance Micro Devices, Fairchild, Harris, ITT, Motorola, Philips, RCA, Raytheon, Signetics, Sescosem, SGS-ATES, Siemens, AEG-Telefunken, Texas Instruments, Teledyne.

Companion volume to book No. BP40 — DIGITAL IC EQUIVALENTS AND PIN CONNECTIONS.

An invaluable addition to the library of all those interested in Electronics, be they amateur or professional.

WORLD WIDE COMPREHENSIVE RADIO VALVE GUIDE BOOKS
1/2/3/4/5   **PRICE: 40p**
Approx. Size: 170 x 115 mm
100 Book 1, Author: W.J. May   ISBN: 0 900162 01 5
121 Book 2, Author: B.B. Babani   ISBN: 0 900162 02 3
143 Book 3, Author: B.B. Babani   ISBN: 0 900162 07 4
157 Book 4, Author: B.B. Babani   ISBN: 0 900162 13 9
178 Book 5, Author: B.B. Babani   ISBN: 0 900162 24 4

These 5 different Valve Data books cover more than 11,000 valves and include all receiving, miniature, sub-miniature, television cathode ray tubes, voltage and current stabilisers, thyratrons, rectifiers, produced in Great Britain, USA, France, Holland, Germany, USSR, Poland, Czechoslovakia, Japan, Australia and all other countries of the world producing electronic valves, thus making the five books world-wide. A complete numerical/alphabetical index is in every book and shows the type of base and the page number where the operating characteristics can be found.

202: HANDBOOK OF INTEGRATED CIRCUITS (IC's) EQUIVALENTS AND SUBSTITUTES

AUTHOR: B. B. BABANI   PRICE: 75p

ISBN: 0 900162 35 X   128 Pages

Approx. Size: 180 x 105 mm

One of the first and most complete integrated circuits (IC's) equivalents and substitutes guide ever published containing full interchangeability data on more than 9,500 integrated circuits with every possible alternative and equivalent clearly shown. Comprehensively covers digital and linear IC's of every type, including those manufactured in Great Britain, USA, Japan, Germany, France, Czechoslovakia, the rest of Europe and all other manufacturing sources. The products of the world's leading makers are listed in this unique book. Most available commercial industrial service and military types are extensively covered.

211: FIRST BOOK OF DIODE CHARACTERISTICS, EQUIVALENTS AND SUBSTITUTES

AUTHOR: B. B. BABANI   PRICE: 95p

ISBN: 0 900162 46 5   160 Pages

Approx. Size: 180 x 105 mm

Full instructions for using this unique book given in 9 languages: English, Deutsch, Francais, Nederlands, Espanol, Italiano, Portugues, Dansk, Svenska. This book includes signal diodes, zener diodes and rectifiers, etc. One of the first complete guides ever published. Shows full interchangeability data and characteristics of many thousands of diodes of all types with every possible alternative. Includes all British, European, Soviet, American, Continental and Far Eastern diodes on the market today. All types of domestic, industrial and military designs extensively covered. More than 25,000 entries.

Please note overleaf is a list of other titles that are available in our range of Radio and Electronic Books.

These should be available from most good Booksellers, Radio Component Dealers and Mail Order Companies.

However, should you experience difficulty in obtaining any title in your area, then please write directly to the publishers enclosing payment to cover the cost of the book plus adequate postage.

If you would like a copy of our LATEST CATALOGUE covering all our publications, then please send a stamped addressed envelope to:

BERNARD BABANI (publishing) LTD
THE GRAMPIANS
SHEPHERDS BUSH ROAD
LONDON W6 7NF
ENGLAND